Liquid Chromatography of Polymers and Related Materials

CHROMATOGRAPHIC SCIENCE

A Series of Monographs

Other Volumes in Preparation

Liquid Chromatography of Polymers and Related Materials

Edited by JACK CAZES

Waters Associates, Inc.
Milford, Massachusetts

MARCEL DEKKER, INC. New York and Basel

Library of Congress Cataloging in Publication Data

International Symposium on Liquid Chromatographic Analysis of Polymers and Related Materials, 1st, Houston, Tex., 1976.
Liquid chromatography of polymers and related materials.

(Chromatographic science ; v. 8)
Includes indexes.
1. Polymers and polymerization--Analysis--Congresses.
2. Liquid chromatography--Congresses. I. Cazes, Jack, 1934- II. Title.
QD139.P6I57 1976 547'.84 77-6373
ISBN 0-8247-6592-3

The International Symposium on Liquid Chromatographic Analysis of Polymers and Related Materials was sponsored by Waters Associates, Inc., Milford, Massachusetts.

MARCEL DEKKER, INC.
270 Madison Avenue, New York, New York 10016

Current printing (last digit):
10 9 8 7 6 5 4 3 2 1

PRINTED IN THE UNITED STATES OF AMERICA

PREFACE

Published in this volume are the proceedings of the International Symposium on Liquid Chromatographic Analysis of Polymers and Related Materials, which was held on October 13-15, 1976 at the Rice-Rittenhouse Hotel in Houston, Texas. Assembled here are thirteen papers covering a range of topics of interest to those involved in fractionation and characterization of macromolecular systems.

Jack Cazes
Editor

CONTRIBUTORS

M. R. AMBLER, Chemical Materials Development, The Goodyear Tire and Rubber Company, Akron, OH 44316

R. W. ASHCRAFT, Development Division, Mason and Hanger-Silas Mason Company, Inc., P.O. Box 647, Amarillo, TX 79177

J. CAZES, Waters Associates, Inc., Maple Street, Milford, MA 01757

R. P. CHARTOFF, Department of Chemical and Nuclear Engineering, University of Cincinnati, Cincinnati, OH 45221

D. J. CRABTREE, Aircraft Division, Northrop Corporation, 3901 West Broadway, Hawthorne, CA 90250

B. L. DAWSON, IBM Research Laboratory, San Jose, CA 95193

E. E. DROTT, Monsanto Textiles Company, P.O. Box 12830, Pensacola, FL 32575

M. Y. HELLMAN, Bell Laboratories, Murray Hill, NJ 07974

D. B. HEWITT, Aircraft Division, Northrop Corporation, 3901 West Broadway, Hawthorne, CA 90250

T. C. HUARD, Waters Associates, Inc., Maple Street, Milford, MA 01757

D. E. JOHNSON, IBM Research Laboratory, San Jose, CA 95193

E. KOHN, Development Division, Mason and Hanger-Silas Mason Company, Inc., P.O. Box 647, Amarillo, TX 79177

B. LIGHTBODY, Research and Development Department, Waters Associates, Inc., Maple Street, Milford, MA 01757

S. K. T. LO, Department of Chemical and Nuclear Engineering, University of Cincinnati, Cincinnati, OH 45221

N. MARTIN, Waters Associates, Inc., Maple Street, Milford, MA 01757

R. D. MATE, Chemical Materials Development, The Goodyear Tire and Rubber Company, Akron, OH 44316

R. D. NUSS, Brunswick Corporation, 4300 Industrial Avenue, Lincoln, NB 68504

A. C. OUANO, IBM Research Laboratory, San Jose, CA 95193

H. QUINN, Research and Development Department, Waters Associates, Inc., Maple Street, Milford, MA 01757

R. L. SAMPSON, Millipore Corporation, Bedford, MA 01730

N. THIMOT, Research and Development Department, Waters Associates, Inc., Maple Street, Milford, MA 01757

R. VIVILECCHIA, Research and Development Department, Waters Associates, Inc., Maple Street, Milford, MA 01757

CONTENTS

Liquid Chromatography of Polymers and Related Materials

CHARACTERIZATION OF P(MMA) AND P(MMA/MAA) BY LOW ANGLE LASER LIGHT SCATTERING PHOTOMETRY, GPC, AND VISCOMETRY

A. C. Ouano
B. L. Dawson
D. E. Johnson

IBM Research Laboratory
San Jose, California

ABSTRACT

Solution properties of poly(methyl methacrylate), P(MMA), and (methyl methacrylate/methacrylic acid) copolymers, P(MMA/MAA), were investigated using low angle laser light scattering (LALLS), gel permeation chromatography (GPC) and viscometry techniques. The molecular weight and compositional ranges studied were from 2.6×10^4 to 3×10^6 and from p(MMA) to a 50/50 P(MMA/MAA) copolymer ratio, respectively.

In tetrahydrofuran (THF) solvent, the exponent of the Mark-Houwink equation was found to vary between 0.73 for P(MMA) to 0.63 for 50/50 copolymer. A plot of the exponent of the Mark-Houwink equation versus the % acid in the copolymer appears to be linear and approaches a value of 0.50 as the % acid approaches 75%. A comparison between the weight average molecular weight ($\bar{M}w$) obtained by LALLS techniques and those obtained by GPC and viscometry using polystyrene calibration standards indicated the P(MMA) and P(MMA/MAA) elute from GPC according to their hydrodynamic size. This allows the estimation of the molecular weight distribution (MWD) from GPC chromatograms using the Universal Calibration method.

INTRODUCTION

With the possible exception of polystyrene and other olefinic polymers, poly(methyl methacrylate) P(MMA) is probably the most studied polymer, for its solution properties have been reported by a number of investigators (1-6). However, relatively little is known about its methacrylic acid copolymers, P(MMA/MAA).

Panov and his co-workers, (7-10) have studied some of the solution properties of these copolymers in N,N-dimethyl formamide (DMF) and ethylene dichloride (EDC) at various molar ratios of the acid and ester comonomers. Panov and Frenkel (7) reported that (MMA/MAA) copolymers in EDC solvent do not follow the Mark-Houwink type relationship between the intrinsic viscosity, $[\eta]$, and molecular weight, M. Their results showed a highly non linear relationship between the log $[\eta]$ and log [M] for P(MMA/MAA) with $M \geq 10^5$ and MAA mole % $\geq$ 10%. Based on these results Panov and Frenkel (7) concluded that this non-linear $[\eta]$-M relationship might also hold true for other solvents. This conclusion implies that the (MMA/MAA) copolymers will not fractionate in the gel permeation chromatograph, GPC, according to their molecular hydrodynamic size, hence making the so called "Universal Calibration" method (11) of estimating the molecular weight distribution from the GPC chromatogram invalid. Thus, a primary purpose of this work was to ascertain the validity of Panov and Frenkel's conclusion for the P(MMA/MAA) in THF solvent. Our other objective was to determine qualitatively the existence of any dependence of the characteristic ratio $C_n = \frac{\langle r^2 \rangle_o}{n\ell^2}$ on the acid molar ratio in P(MMA/MAA).

EXPERIMENTAL

Synthesis

Most of the copolymers were prepared in THF (redistilled from $LiAlH_4$) and the remainder in toluene (redistilled from Na). For a typical synthesis, a heavy walled, 175 ml pressure bottle was oven dried at 170°C and cooled under dry N_2. The addition of reagents was carried out under a blanket of dry N_2 and subsequently the bottle sealed via a Teflon sleeve fitted over the bottle's stopper. To the reaction bottle were added 100 ml THF, 15 ml(14.1g, 0.14 mole) MMA, which had been passed through neutral $A\ell_2O_3$, 12 ml (12.2g, 0.14 mole) MAA, vacuum redistilled, and 0.1g of t-butyl hydroperoxide. The solution was heated to 70°C for 48 hrs. and the product recovered by pouring the viscous solution into excess n-hexane. The precipitate was washed several times with n-hexane and dried in a vacuum oven at 60°C for 18 hours. The acid content was found to be 48%. For copolymers with other acid contents, the monomer feeds were adjusted accordingly. For copolymers with M_w's less than 100K, bromotrichloromethane was used as the M_w regulator. To prepare the fully methylated samples, the copolymers were dissolved in $CHCl_3$-MeOH mixtures and treated with an excess of diazomethane in ether.

Compositional Analysis

The copolymers were analyzed either by titrating them with standard base or by using ^{13}C-NMR. To titrate the copolymers, samples were analytically weighed into a beaker and dissolved in an EtOH-H_2O mixture. The solution was titrated to a phenolphthalein endpoint with standardized 0.15 N aqueous KOH.

For ^{13}C-NMR analyses, 0.3g samples were dissolved or swelled in ca. 2g of a 50/50 mixture of pyridine and pyridine-d_5, both compounds being directly weight into a 10mm NMR tube. The ratios of ester to acid were determined by comparing the integrals of the carbonyl peaks from 175 to 185ppm relative to TMS.

M and [η] Measurement

All solutions in this study were prepared in THF (distilled in glass grade, Burdick and Johnson) stabilized with 0.1% Ionol. The weight average molecular weights (M_w) were measured using the recently developed low angle laser light scattering (LALLS) photometer (13-15).

The Rayleigh factors of the different samples were measured at an angle of 4° scattering over a range of concentrations. The M_w and the second virial coefficient A_2 were calculated by linear regression analysis from the intercept and the slope of KC/R_θ (K is the polymer constant and R_θ is the Rayleigh factor) versus concentration (C) respectively. The C versus the KC/R_θ plot for all samples had a linear regression correlation factor of better than 0.99 (1 is perfect linear correlation).

The solution viscosities were measured by an automatic viscounter (FICA VISCOMATIC) over a concentration range, and were adjusted for the different molecular weight samples such that the efflux times through the capillary were between 20 to 40 secs. The intrinsic viscosities were obtained from the intercepts of both the ($\eta sp/C$) and ln (ηr)/C versus C plots. Both plots were linear within the concentration ranges of the measurements.

The GPC analyses were carried out in a modified and automated (data acquisition and reduction are computerized) Waters Associates Model 200. The modifications on the GPC were described elsewhere (14). Five 4 ft fractionating columns with permeability limits ranging from 10^3 Å to 5×10^6 Å (Water's Associates designation) and column efficiencies of from 800 to over 1500 plates/ft were used. The calibration standards used were narrow molecular weight distribution (MWD) polystyrenes supplied by Pressure Chemical Co. and the National Bureau of Standards (NBS-705). No band broadening corrections were made in calculating the the value of M_w from the GPC data.

Results and Discussions

The $[\eta]$-M relationships for both P(MMA) and a 25 mole % MAA copolymer in Figure 1 show a good linear correlation. In fact, linear regression analyses of the $[\eta]$-M data exhibited correlation factors of 0.98 or higher for all samples. This linearity held throughout the molecular weights (2.6×10^4 to 3×10^6) and MAA content ranges (0 to 50%) studied. Thus it appears that Panov's and Frenkel's conclusion does not hold for P(MMA/MAA) in THF solvent.

A linear regression analysis of the log $[\eta]$-log M data gave the Mark-Houwink coefficients K and exponents α. The value of α was found to decrease with increasing acid content of the P(MMA/MAA). A plot of α and K versus the acid content of the copolymer is exhibited in Figure 2. The relationship between α and mole % MAA appears to be linear and extrapolation to zero acid content (PMMA) shows an α value of 0.75. Extrapolating the value of α to 0.50, indicates the theta condition at room temperature for P(MMA/MAA) in THF to be at 75 mole % MAA content. The second virial coefficient A_2, also appears to decrease with increasing acid content from 6.0×10^{-4} for P(MMA), to 3.4×10^{-4} for a 50 mole % P(MMA/MAA) with equivalent molecular weights.

A comparison of the M_w data obtained by the LALLS and GPC - "Universal Calibration" techniques shown in Table I, indicate good agreement. The M_w values for the linear portion of the GPC calibration (8×10^5 to 5×10^4) curve, agree with the light scattering values within 15%. These results indicate that (MMA/MAA) copolymers (up to 50 mole % MAA) in THF at room temperature fractionate in GPC according to their molecular hydrodynamic size in solution.

The characteristic ratios C_n were calculated from the K_θ in THF (a good solvent) using both the Fox-Flory (F-F) (16)

$$[\eta]^{2/3}\, M^{-1/3} = K_\theta^{\,2/3} + B_1\, (M/[\eta]) \qquad (1)$$

and the Stockmayer-Fixman (S-F) (17)

$$[\eta]M^{-1/2} = K_\theta + B_2\, M^{1/2} \qquad (2)$$

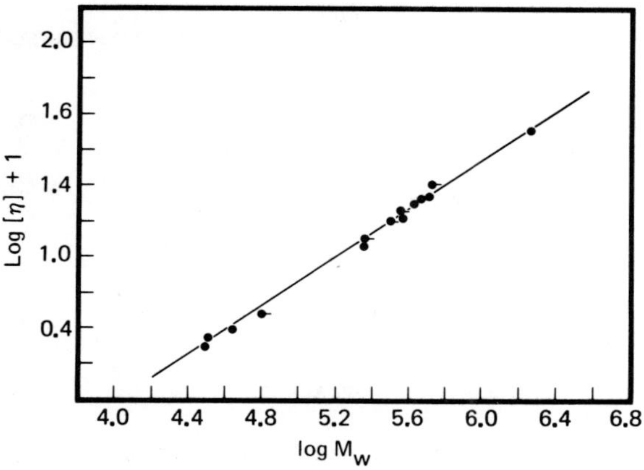

Figure 1. A logarithmic plot between [η] and M for both P(MMA) ●- and 25% MAA copolymer ● .

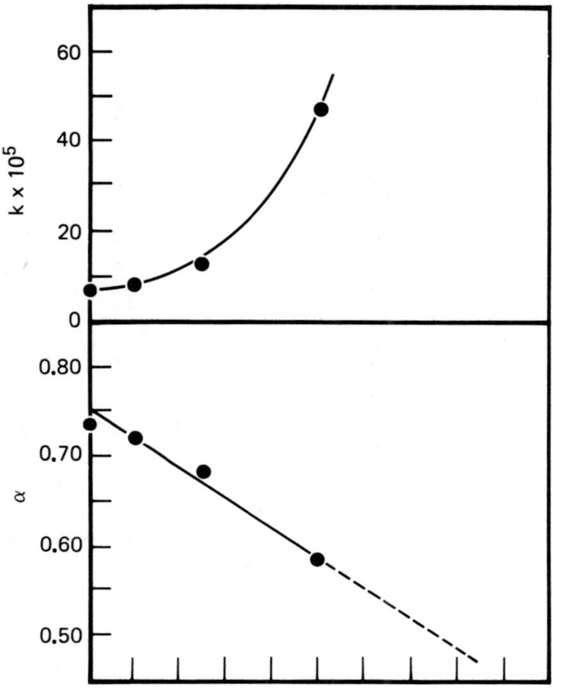

Figure 2. The Mark-Houwink exponent α and coefficient K plotted against percent MAA in the copolymer.

TABLE I

Comparison of the M_w Obtained by GPC-Universal Calibration and Light Scattering Techniques

Sample ID	Mole % MAA	M_w	
		GPC*×10^{-3}	LALLS×10^{-3}
88-2M	0	370	345
88-3M	0	510	513
103-2M	0	650	667
96-2M	0	765	830
108-2M	0	1100	1250
10-3	10	56	50
21-1	10	58	60
10-6	10	82	81
108-2	10	1230	1250
16-3	25	56	49
10-2	25	57	50
10-5	25	79	68
88-2	25	350	357
88-3	25	545	546
04-49	25	576	546
103-2	25	625	667
02-4	25	670	690
10-4	50	32	26
10-1	50	50	43
96-2	50	805	1042

*Average of Triplicate values.

extrapolation techniques. Figure 3 shows a plot of the $[\eta]M^{-1/2}$ versus $M^{1/2}$ for P(MMA) and a 25 mole % acid copolymer. A linear regression analysis of the $[\eta]$-M data following both equations (1) and (2) gave correlation factors of greater than 0.95 for all molecular weight and acid content ranges investigated. The root mean square end to end distances $\langle r^2 \rangle_o$ were computed from

$$\langle r^2 \rangle_o = [K_\theta/\Phi]^{2/3} M \qquad (3)$$

where Φ is the Flory constant.

The extrapolated values of K_θ and the calculated values of $\langle r^2 \rangle_o$ and C_n for the different molar contents of MAA are presented in Table II. The K_θ obtained by Chinai and Valles (2) for P(MMA) (the tacticity of which was not reported) in a theta solvent was 5.9×10^{-4}. This agrees reasonably well with the average value of 6.0×10^{-4} we obtained using the Fox-Flory

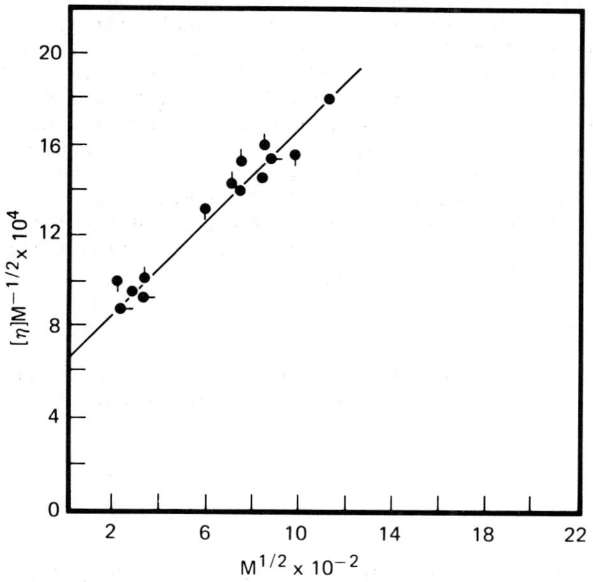

Figure 3. The Stockmayer-Fixman extrapolation technique for estimating K_θ. P(MMA) ● , 10% MAA ● , 25% MAA ●, 50% MAA ● .

TABLE II

Characteristic Ratios of P(MMA/MAA) with Various MAA Molar Ratios

Mole % MAA	$K_\theta \times 10^4$			C_n		
	S-F	F-F	AV•VALUES	S-F	F-F	AV•VALUES
0	7.0	4.9	6.0	9.2	7.3	8.2
10	6.6	4.2	5.4	8.8	6.5	7.6
25	7.2	6.6	6.9	9.4	8.8	9.1
50	9.2	8.3	8.8	10.5	10.2	10.4

and the Stockmayer-Fixman extrapolation techniques. Sundararajan and Flory (1) calculated from configurational characteristics the C_n values of P(MMA) with various tacticities. The average C_n value for 60% syndiotactic P(MMA) of 8.2 reported in Table II agrees quite well with the 8.5 value of C_n reported by Sundararajan and Flory (1) for predomenantly syndiotactic P(MMA). The dependence of C_n on the mole % of MAA in the copolymer appears to be very small if it exists. The small increase in the C_n value with increasing mole % MAA could be due to a small variation in the taciticity of the samples, and the natural scatter of the data expected from the extrapolation techniques used to estimate the value of K_θ.

ACKNOWLEDGMENTS

The authors wish to thank L. A. Pederson for his assistance in synthesizing these polymers and T. T. Horikawa for the ^{13}C-NMR Analyses.

REFERENCES

1. P. R. Sundararajan and Flory, J. Am. Chem. Soc. 96, 5026 (1974).

2. S. N. Chinai and R. J. Valles, J. Polymer Sci. 39, 363 (1959).

3. P. Vasudevan and M. Santappa, J. Polymer Sci. Part A-2, 9, 483 (1971).

4. S. Krause and E. Cohn-Ginsberg, J. Phys. Chem. 67, 1479 (1963).

5. G. V. Schulz and R. Kirste, Z. Phys. Chem. (Frankfurt Am Main) 30, 171 (1961).

6. I. Sakurada, A. Nakajima, O. Yoshijaki, and Ko Nakamae, Kolloid - Z. 186, 41 (1962).

7. Y. N. Panov and S. Y. Frenkel, European Polymer J. 8, 1067 (1972).

8. Y. N. Panov and S. Y. Frenkel, Vysokomolek, Soldin 4A, 116 (1962).

9. Y. N. Panov and S. Y. Frenkel, Vysokomolek, Soldin 9A, 937 (1967).

10. Y. N. Panov, K. E. Norbek and S. Y. Frenkel, Vysokomolek, Soldin 6A 47 (1964).

11. Z. Grubisic, P. Rempp, and H. Benoit, J. Polymer Sci. B5, 753 (1967).

12. W. Kaye, Anal. Chem. 45, 221A (1973).

13. A. C. Ouano and W. Kaye, J. Polymer Sci., Chem. Ed. 12, 1151 (1974).

14. A. C. Ouano, J. of Chromatography, In Press.

15. W. Kaye and A. J. Havlik, Appl. Optics 12, 541 (1973).

16. P. J. Flory and T. G. Fox, J. Am. Chem. Soc. 73, 1904 (1951).

17. W. H. Stockmayer and M. Fixman, J. Polym. Sci. Part C, 1, 137 (1963).

EXCLUSION CHROMATOGRAPHY OF HYDROPHILIC AND OTHER POLYMERS

R. Vivilecchia, B. Lightbody, N. Thimot, and H. Quinn

Research and Development Department
Waters Associates, Inc.
Milford, Massachusetts

INTRODUCTION

During the last five years, high speed, high resolution liquid chromatography has been realized by the development of high efficiency columns packed with small particle (<10 micron) supports.[1-6] The technique of exclusion chromatography (EC), developed in the mid-1960's,[7,8] required long columns to achieve sufficient resolution; consequently, long analysis times were observed. With the development of small particle supports for EC, analysis times of less than twenty minutes are possible.[2,6,9-12] The very high efficiencies resulting from the use of porous particles of less than ten micron are due to the shorter distances the solutes molecules must diffuse within the packed bed. The solute molecules undergo more rapid mass transfer between the mobile phase and the stagnant mobile phase within the pores of the particle. Because of the inherent slower diffusivities of polymeric species in liquids, reduction in particle size greatly improves resolution and reduces analysis times. Such results have been primarily restricted to the non-aqueous soluble polymers.

Hydrophilic polymers offer unique problems which have hindered their analysis by high speed EC. In general, the soft organic gels such as cross-linked dextrans (sephadex) and the poly-acrylamides[14] have been most frequently used. Such gels, because of their poor mechanical stability under pressure, cannot be used as small particles. More recently, silica which has excellent mechanical stability has seen some use for hydrophilic polymers.[15-20] The major disadvantage is adsorption effects upon its polar surface. The surface of silica contains polar, acidic silanol groups (Si-O-H) which can behave as cation-exchange sites and thus lead to strong adsorption of positively changed solutes. Regnier et al.[21] has modified the surface of controlled pore glass with bonded organic polymer layers to prevent interaction with the polar silica surface.

The objective of this work was to develop a support having the following characteristics: rigid particles having high pressure stability independent of the pore size or particle size (2) no shrinking or swelling properties with changing pH, ionic strength or other mobile phase change (3) non-ionic surface to prevent ion-exchange interaction (4) hydrolytically stable surface over a wide pH range (5) closely controllable narrow pore size distributions (6) a hydrophilic, water wettable, surface.

EXPERIMENTAL

APPARATUS

The Waters Associates Model ALC/GPC-244 liquid chromatograph (Waters Assoc. Inc, Milford, MA) was used throughout this work. Columns are constructed with specially designed low dead volume endfittings to minimize extra column band spreading.

CHEMICALS

The chromatographic solvents were distilled-in-glass from Burdick and Jackson Labs (Muskegan, Mich.). The narrow molecular weight polystyrene standards are from Waters Associates. Other polymeric samples were obtained from various suppliers and manufacturers.

A silica based material having an average particle size of 8-10 micron and having controlled, narrow pore size distributions with average pore sizes of 125Å, 300Å, 500Å and 1000Å was produced by a proprietary process. The surface was modified by reacting the surface silanol groups with an organosilane having ether functional groups. The materials were packed into 3.9mm i.d. x 300mm length 316 stainless steel tubes by a modified viscosity technique.[4] These materials will be referred to as μBondagel E. The μStyragel columns (Waters Assoc. Inc., Milford, MA) were 7.8mm i.d. x 300mm length.

RESULTS AND DISCUSSION

μBONDAGEL E CHARACTERIZATION

μBondagel E is a silica based chromatographic packing whose surface has been modified by reaction of surface silanol groups with an ether functional organosilane. A monomolecular layer of organic groups were produced according to the following reaction:

$$-Si-O-H + Cl-Si-R \rightarrow -Si-O-Si-R + HCl$$

A monolayer of organo groups was chosen over polymeric layers, since a drastic decrease in observed pore volume can occur with polymeric layers by blockage of pore openings. Such blockage becomes especially important with pore sizes less than about 300Å.

Figure 1 shows the polystyrene calibration curves of μBondagel E columns. The 1000Å material shows an exclusion limit of about 2 million molecular weight. For the low molecular weight (<1000mw) range, attempts at producing a bonded material with an average pore size of 60Å, led to a fifty percent reduction in pore volume. Thus even monomolecular organosilane coverage leads to loss in pore volume. The organosilane groups can be visualized like bristles projecting out from the surface of the pores. Consequently as the pore opening decreases in size, the bonded organic groups more effectively block the ability of solutes to diffuse into the pore. With average pore sizes greater than 125Å no loss in available pore volume was noted. Thus figure 1 shows that

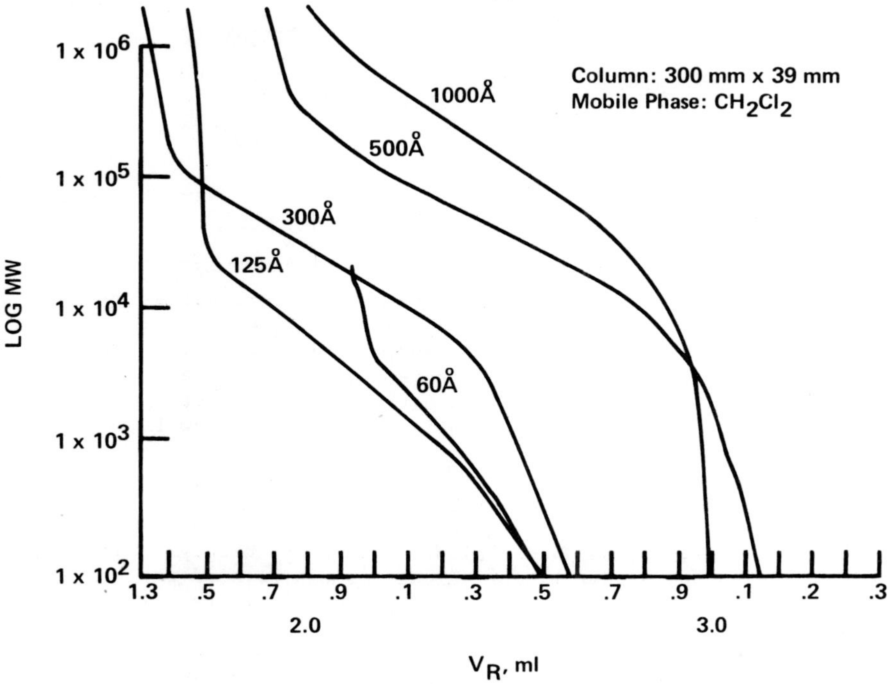

Figure 1. μBondagel E Polystyrene Calibration Curves.

the 60Å material shows no advantage over columns packed with μBondagel E 125Å. Table 1 compares μStyragel and μBondagel E pore sizes. The μStyragel size nomenclature refers to the exclusion limit of a polystyrene standard in terms of its extended chain length in angstroms. The μBondagel nomenclature refers to the experimentally determined average pore size in angstroms. Currently, silica based packings for exclusion chromatography have the greatest deficiencies in resolving power for the separation of small molecules. This deficiency is primarily due to the difficulty in producing high porosity, mechanically stable silica particles with pore sizes less than 60Å. In general the smaller the average pore size the smaller the pore volume.

TABLE 1

Comparison of μStyragel and μBondagel E Pore Sizes

μSTYRAGEL	μBONDAGEL
10^6Å	—
10^5	1000Å
10^4	500
10^3	300
500	125
100	

On the other hand, μStyragel 100Å attains high pore volume by swelling of the gel by the mobile phase. Excellent resolution can be attained with molecules of <600mw by diffusion into the highly swollen gel network.[6,7] However, with swollen gels mechanical stability of the chromatographic bed is sacrificed. Consequently, low flow rates and only solvents which swell the gel must be used.

Since exclusion chromatography (EC) is a diffusion-controlled separation process, reduction in particle size promotes more rapid mass transfer, and hence higher column efficiency. Shorter column length and higher linear velocities can be used to produce faster and more efficient separations. Figure 2 shows an example of a high-speed separation of polystyrene standards from 2 million to 600 molecular weight on a set of μBondagel columns in 6 minutes. In order to get such high resolution, it is necessary to pack columns which give low peak asymmetry. The tangent method of measuring plate count was found to be unsuitable in that it gave unrealistically high plate counts. Instead, column efficiencies were calculated at 4.4% of peak

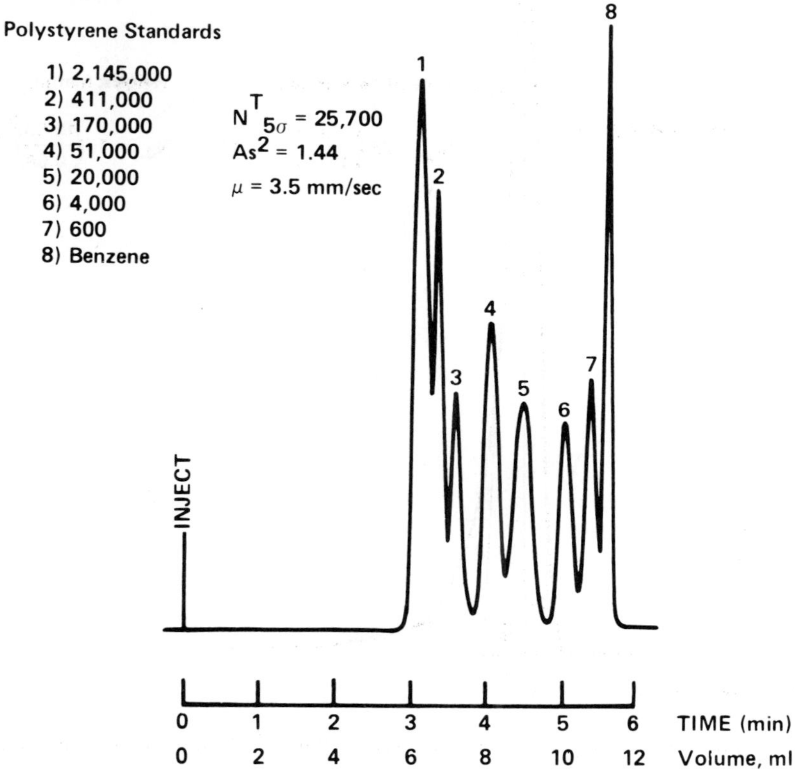

Figure 2. Separation of Polystyrene Standards on μBondagel E Column Set (125, 300, 500, 1000). Mobile Phase: CH_2Cl_2, 2.0 ml/min, ΔP = 2500 psi.

height (5σ value) according to equation (2).

$$N_{5\sigma}=25\left(\frac{V_R}{W_{4.4}}\right)^2 \quad (2)$$

In addition, columns were determined acceptable only if peak asymmetry was ≤2 as calculated by equation (3)

$$As^2=\left(\frac{T}{F}\right)^2 \quad (3)$$

where T is the distance from the tailing end of the peak to the perpendicular from the peak maximum to the baseline and F is the distance from the front end of the peak to the perpendicular. Figure 3 shows examples of these calculations on a poor column. Note the unrealistically high plate counts determined by the tangent method for asymmetric peaks. As peak shape approaches a gaussian profile the differences between the plate counts calculated by both methods diminishes.

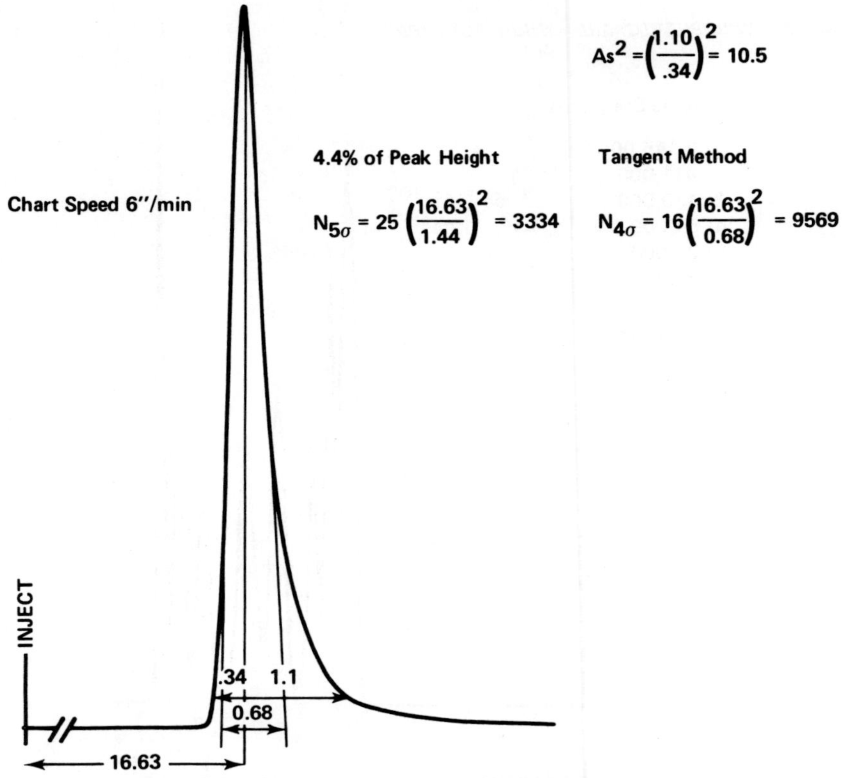

Figure 3. Calculations of Plate Count$^{(N)}$ and Peak Asymmetry (As^2).

TABLE 2

EFFICIENCIES OF POLYSTYRENE STANDARDS

$\overline{d_p}$, µm	µSTYRAGEL™ 10^5 14		µBONDAGEL™ E 10^3 8	
	$N^*_{5\sigma}$	As^2	$N^*_{5\sigma}$	As^2
411 K	280	1.9	1086	1.3
173 K	590	1.4	1465	1.9
111 K	707	1.2	1714	1.4
Benzene	6325	1.1	10,300	1.6

MOBILE PHASE: METHYLENE CHLORIDE
LINEAR VELOCITY: 0.8 mm/sec.

*CORRECTED FOR POLYDISPERSITY

Table 2 shows plate counts and asymmetries of polystyrene standards on a μBondagel and a μStyragel column. Note for these well packed columns the asymmetry values for polymeric standards are also good. Columns which give high asymmetry, will give large errors in molecular weight calculations. Since peak asymmetry may not be independent of molecular weight, accurate correction of elution data may be very difficult. The much higher plate counts observed with the μBondagel column is due to the smaller average particle size. Note that the plate count increases with decreasing molecular weight which agrees with the greater diffusivity as molecular weight is decreased. Figures 4 and 5 shows the increase in resolution that can be obtained by reduction in particle size.

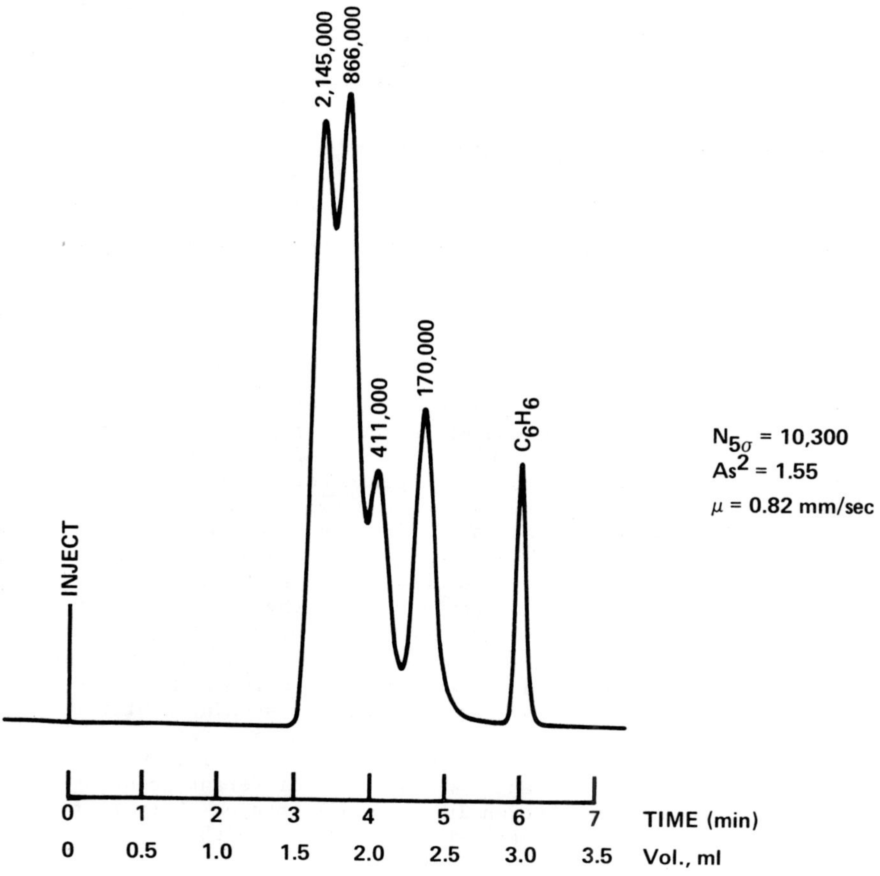

Figure 4. Separation of some Polystyrene Standards on μBondagel E 1000Å. Mobile Phase: CH_2Cl_2, 0.5 ml/min.

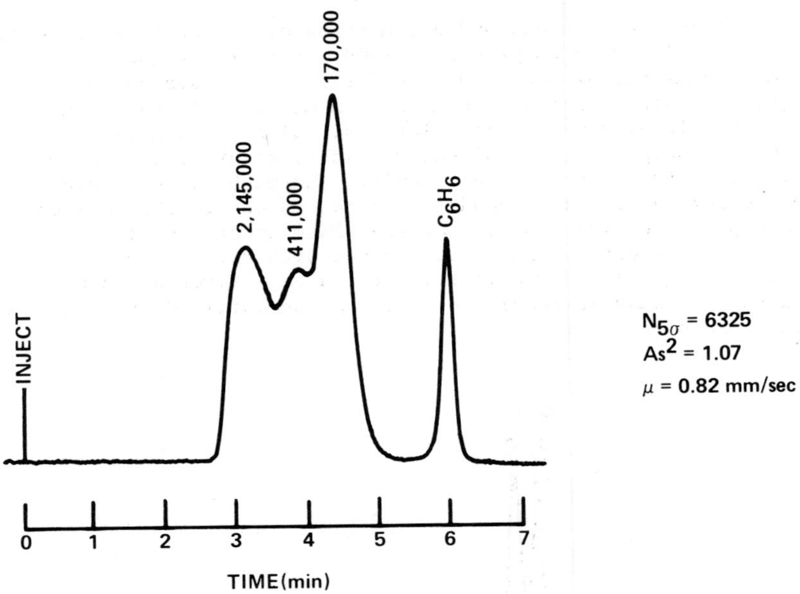

Figure 5. Separation of some Polystyrene Standards on μStyragel 10^5Å. Mobile Phase: CH_2Cl_2, 2.0 ml/min.

μBONDAGEL E LINEAR

Since resolution in EC is a function of the slope of the calibration curve, phase ratio and the plate count, reproducing data between laboratories can be a difficult matter because of column variability in calibration curve and plate count. Variability in columns becomes most critical when columns with different calibration curves are placed in series. In order to minimize this problem, columns with linear calibration curves over the range of 10^6-10^3mw range were produced. With such columns the column length need only be increased to give the desired resolution. Figure 6 shows the calibration curves of polystyrenes and dextrans on four columns in series. Figure 7 shows the separation of some polystyrene standards on this bank of columns. Compare with figure 8 which shows the same polystyrene standards at the same linear velocity on series of columns with different porosities. Note that the linear set gives more or less equal resolution over the whole range, whereas the different porosity columns show greater resolution in the middle range. Very fast separations with high resolution can be achieved with single lengths of columns as is seen in figure 9.

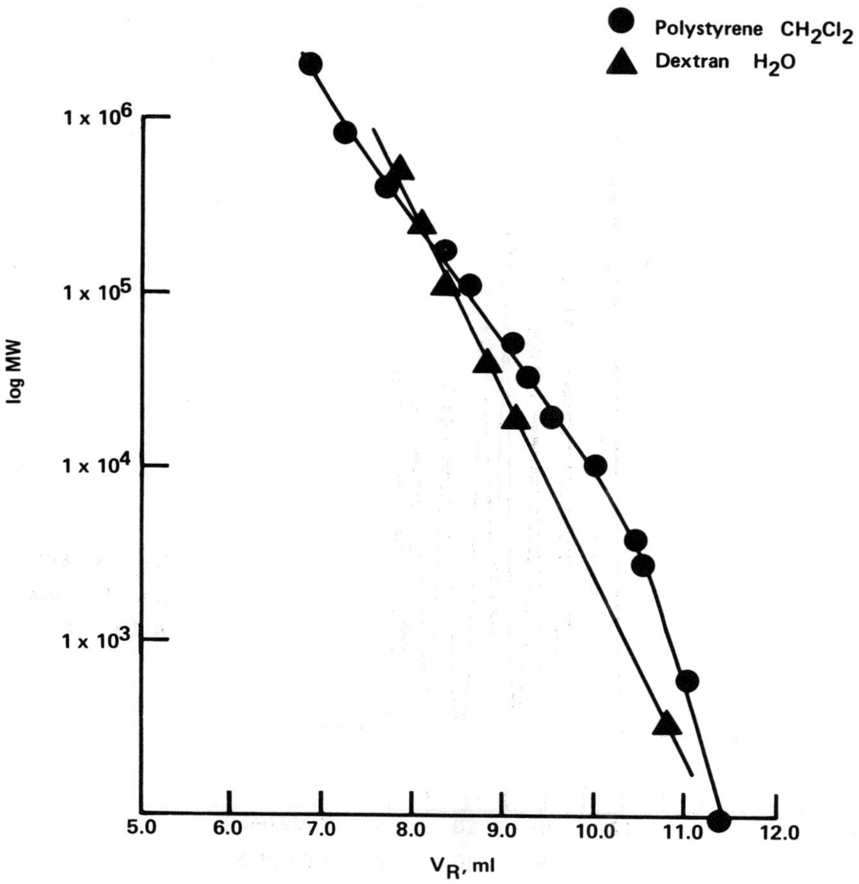

Figure 6. Calibration Curves of μBondagel™ E Linear. Column Dimensions: 4-300mm x 3.9mm (I.D.).

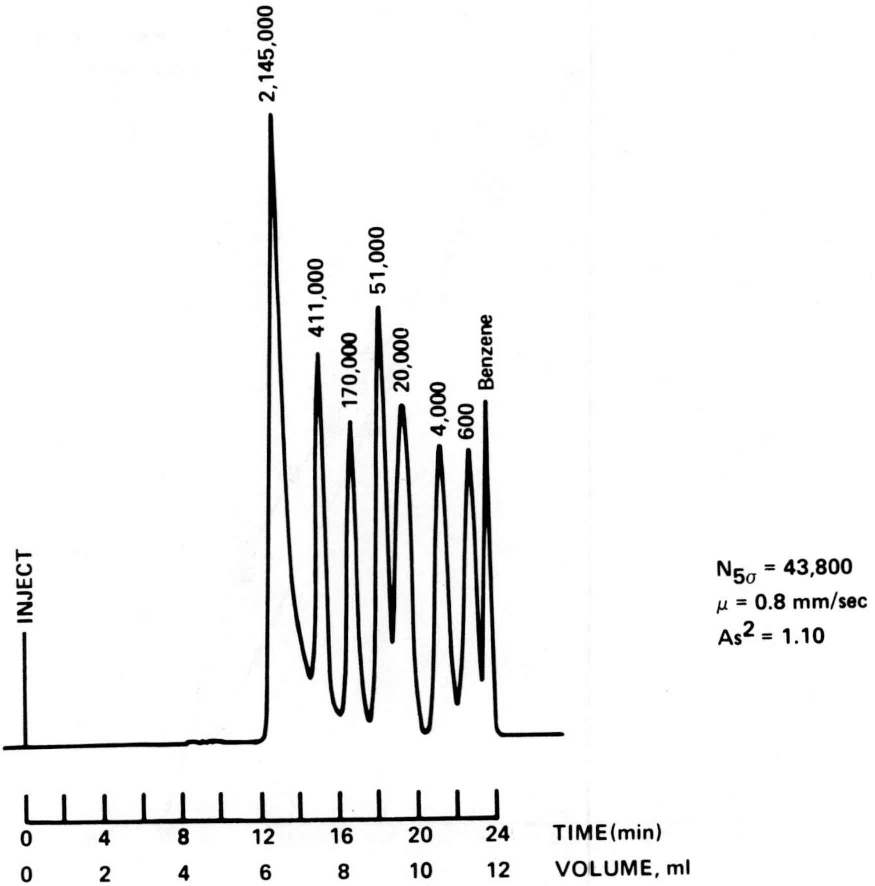

Figure 7. Separation of Polystyrene Standards on μBondagel E Linear. Mobile Phase: CH_2Cl_2, 0.5 ml/min, ΔP=600 psi. Column Dimensions: 4-300mm x 3.9mm (I.D.).

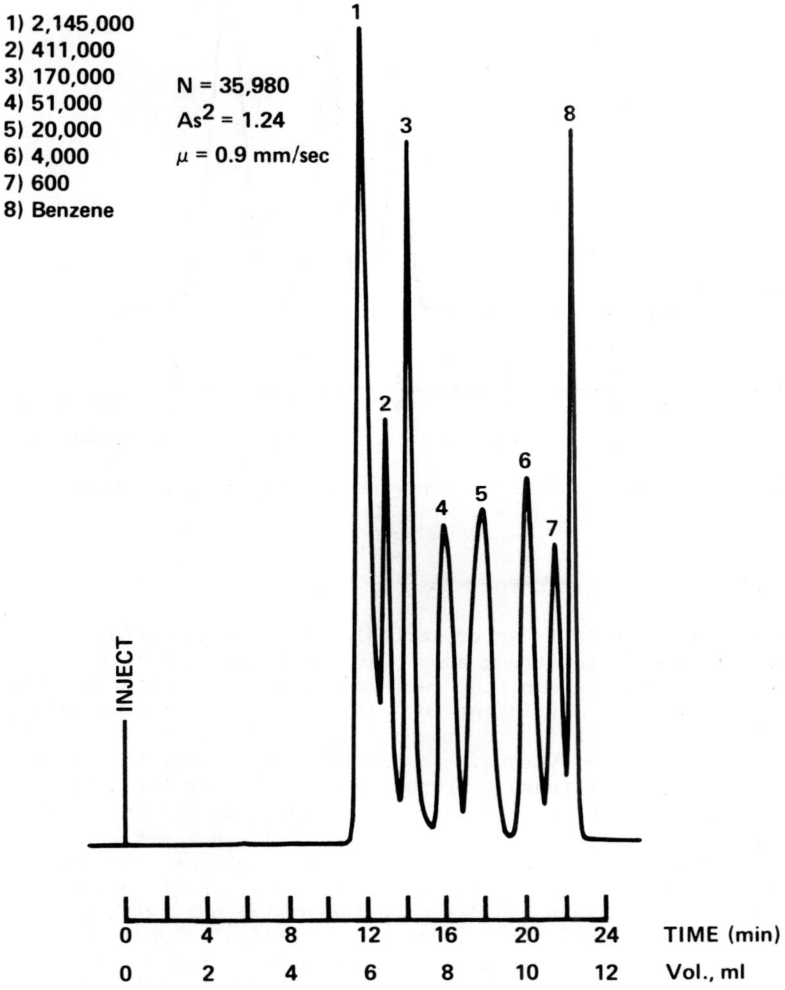

Figure 8. Separation of Polystyrene Standards on μBondagel E column set (125, 300, 500, 1000) Mobile Phase: CH_2Cl_2, 0.5 ml/min, ΔP=650 psi.

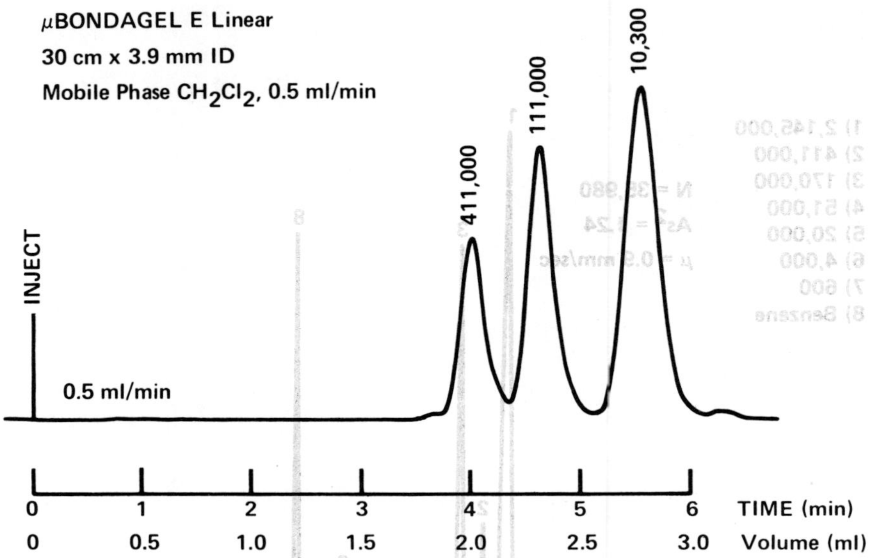

Figure 9. Separation of Polystyrene Standards on μBondagel E Linear.

SEPARATION OF HYDROPHILIC POLYMERS

Although silica itself is water wettable, the strong surface polarity gives rise to strong adsorption with many water soluble polymers. For this reason, the surface was modified with an ether functional silane of sufficient polarity to give water wettability but of low enough polarity to prevent strong adsorption in aqueous or nonaqueous systems. In addition, the ether function-ability shows no ion-exchange properties and is hydrolytically stable. Polyethers have been used successively in preventing the adsorption of proteins on silica surfaces.[18,19] However, unmodified silica has solubility in aqueous solutions. At pH7 and room temperature, silica has a solubility of about 100ppm. The solubility increases rapidly with increasing pH. With the constant flow of aqueous solutions through silica columns of small particles, bed stability because of dissolution can be a serious problem. Reaction of the surface with hydrolytically stable organosilanes reduces the rate of dissolution and thus increases column stability.

Calibration curves of dextrans in water are shown in figures 6 and 10. Figure 11 shows the separation of sulfonated polystyrenes obtained from three different manufactures. Differences in molecular weight distributions are readily apparent. Figure 12 illustrates the separation of a used polyvinylalcohol sample. Other water soluble polymers such as polyacrylic acids,

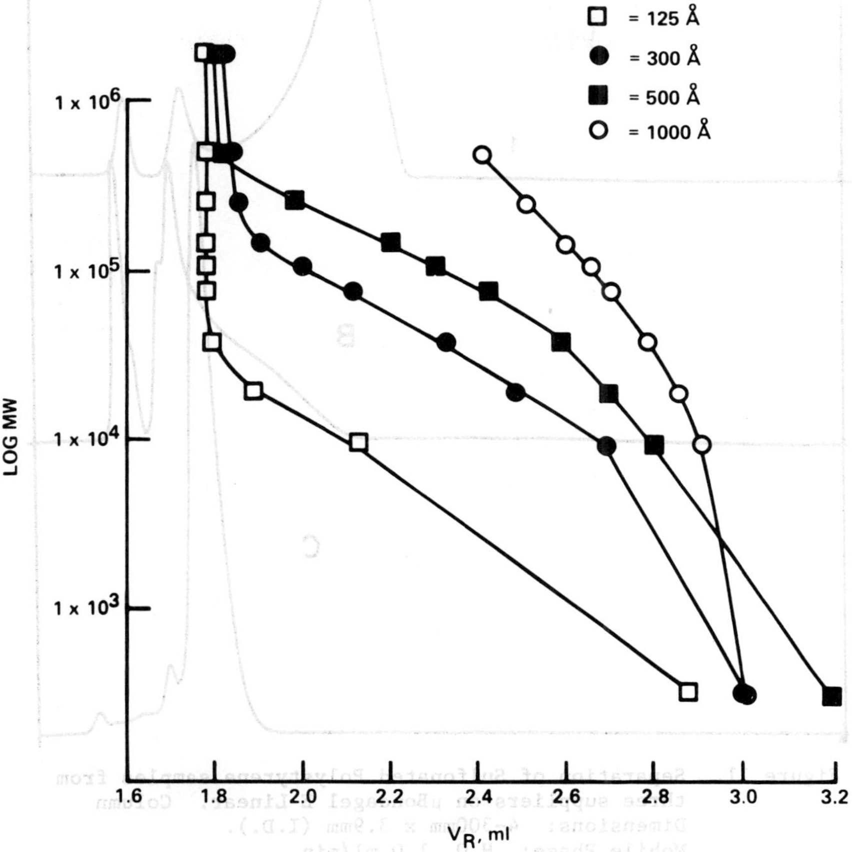

Figure 10. Dextran Calibration of the μBondagel E column set. Mobile Phase: H_2O, 1.0 ml/min.

natural gums, and heparin have been successively chromatographed without adsorption. Proteins on the other hand have shown some adsorption problems. The adsorption was found to be caused by hydrophobic bonding between the proteins and the surface. Increasing salt concentration lead to increased retention which is characteristic of hydrophobic chromatography.[22,23] In addition, changing pH had little effect upon retention. The addition of 0.2-1% sodium dodecylsulfate to the mobile phase gave quantitative recovery of protein as determined by the Lowry test.[24] Figure 13 shows a human plasma profile with possible peak identifications. Apparently, the bonded ether layer although water wettable shows weak hydrophobic character. The addition of ethylene glycol to buffers also gave quantitative elution of proteins from packed columns.

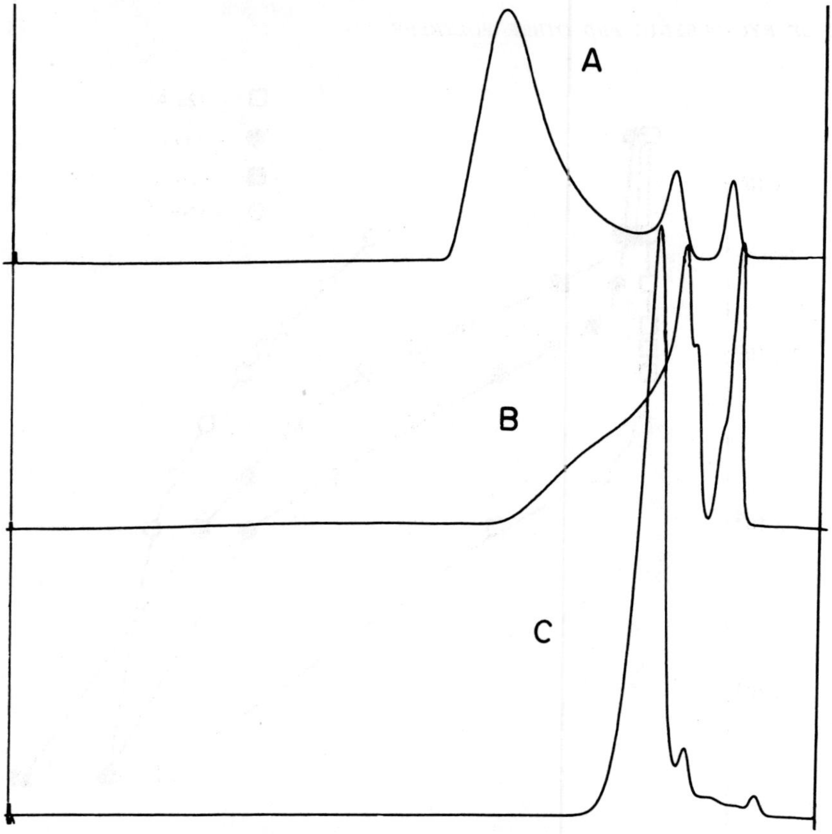

Figure 11. Separation of Sulfonated Polystyrene samples from three suppliers on μBondagel E Linear. Column Dimensions: 4-300mm x 3.9mm (I.D.). Mobile Phase: H_2O, 1.0 ml/min. Chart Speed: 0.75 in/min.

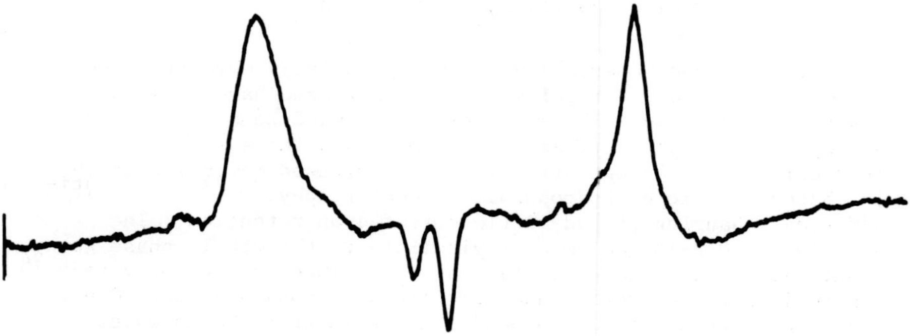

Figure 12. Separation of a used Polyvinylalcohol sample. Column: μBondagel E Linear 2-300 mm x 3.9mm (I.D.) Mobile Phase: Water, 1% Sodium Dodecylsulfate, 1.0 ml/min.

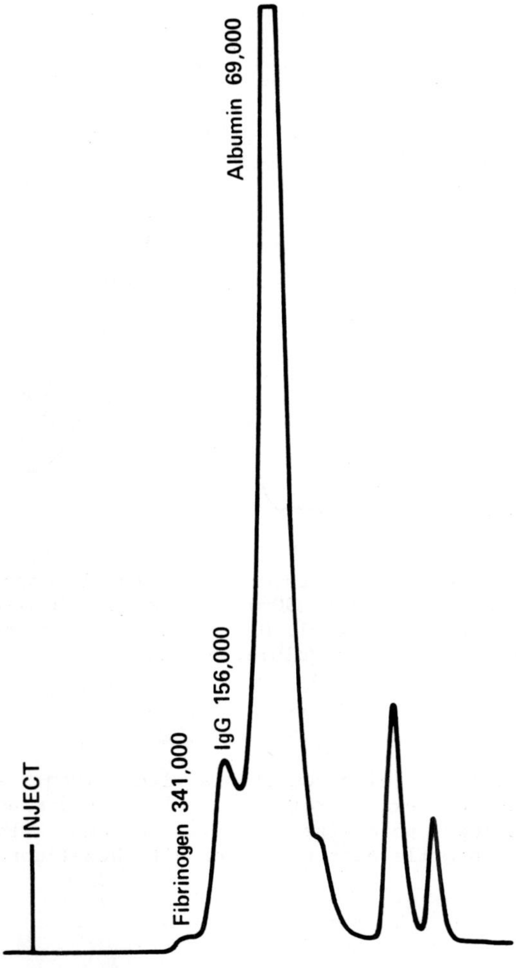

Figure 13. Separation of a Diluted Human Blood Plasma sample on μBondagel E (2-1000Å, 1-500Å, 1-125Å). Mobile Phase: 0.5m Trizma Acetate ph7.4 with 1.0% Sodium dodecylsulfate, 0.3 ml/min. Chart Speed 12 in/hr. Detection UV at 280nm.

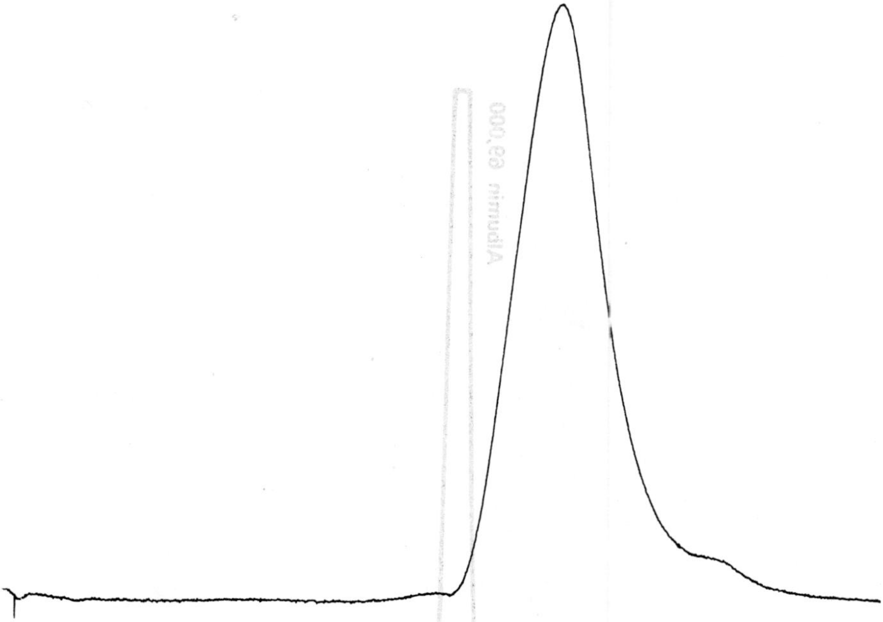

Figure 14 Separation of a Polyester sample on μBondagel E Linear Column Dimension: 2-300mm x 3.9mm I.D. Mobile Phase: Hexafluoroisopropanol, 0.7 ml/min. Chart Speed 2.5 cm/min. Injection Volume: 55 ml.

This packing material is also compatible with solvents such as dimethylformamide and hexafluoroisopropanol which are not compatible with the lower pore size μStyragel columns. For example, figure 14 shows a profile of a polyester with hexafluoroisopropanol as the mobile phase.

CONCLUSION

Silica based packings with bonded phases can show universal compatibility of mobile phases from hexane to water. High speed separation of polymers (6 minutes or less) are possible with well packed volumes having particle sizes less than about 10 micron.

REFERENCES

1. J.F.K. Huber, J. Chromatogr. Sci., 7, 85(1969).
2. J.J. Kirkland, Ibid 10, 593(1972).
3. R.E. Majors, Anal. Chem., 44, 1722(1972).
4. J. Asshauer and I. Halasz, J. Chromatog. Sci., 12, 139(1974).
5. R.M. Cassidy, D.S. LeGay and R.W. Frei, Anal. Chem., 46, 340(1974).
6. R.V. Vivilecchia, R.L. Cotter, R.J. Limpert, N.Z. Thimot, and J.N. Little, J. Chromatog., 99, 407(1974).
7. J.C. Moore, J. Polym. Sci., Part A-Z, 835(1964).
8. D.D. Bly, in B. Caroll (Editor), Physical Methods in Macromolecular Chemistry, Vol. 2, Marcel Dekker, New York, 1972, Chap. 1, pp. 1-89.
9. R.J. Limpert, R. L. Cotter and W.A. Dark, Amer. Lab, 6, No. 5, 63(1974).
10. K. Unger and R. Kern, J. Chromatog., 122, 345(1976).
11. K. Unger, R. Kern, M.C. Ninou and K. Krebs, Ibid. 99, 435(1974).
12. Y. Kato, S. Kido, H. Watanabe, M. Yamamoto, T. Hashimoto, J. Applied Polymer Sci., 19, 629(1975).
13. J. Porath and P. Flodin, Nature, 183, 1657(1959).
14. S. Raymond and Y.J. Wang, Anal. Biochem., 1, 391(1960).
15. W. Haller, Nature, 206, 4985(1965).
16. A.J. deVries, M. LePage, R. Beau, and C.L. Guillemin, Anal. Chem., 39, 935(1967).
17. Cantow, M.J.R., and Johnsow, J. Appl. Polymer Sci., 11, 1851(1967).
18. R.C. Collins, and W. Haller, Anal. Biochem., 54, 47(1975).
19. G.L. Hawk, J.A. Cameron and L.B. Dufault, Prep. Biochem., 2(2), 193(1972).
20. C.W. Hiat, A. Shelokov, E.J. Rosenthal, J.M. Galimor, J. Chromatog., 56, 362(1971).
21. S. Chang, K. Gooding and F. Regnier, J. Chromatog., 120, 321(1976).
22. M. Mevarech, W. Leicht and M.M. Weber, J. Biochem., 15(11), 2883(1976).
23. S. Hjerten, J. Chromatog., 87, 325(1973).
24. O.H. Lowry, N.J. Rosebrough, A.L. Farr and R. J. Randall, J. Biol. Chem., 193, 265(1951).

INTRINSIC VISCOSITY BY GEL PERMEATION CHROMATOGRAPHY: METHOD AND APPLICATION

M. Y. Hellman

Bell Laboratories
Murray Hill, New Jersey

ABSTRACT

Conventional viscometry and G.P.C. are used in obtaining intrinsic viscosities. Both methods are discussed and compared for accuracy. Combining both methods leads to obtaining an absolute calibration curve using only one well characterized standard. Application to the study of melt flow behavior of polycarbonate and branching of polyethylene is discussed.

INTRODUCTION

Gel permeation chromatography (G.P.C.) has demonstrated certain advantages over the tedious and time-consuming measurements of light scattering, osmometry and viscometry. From the G.P.C. trace, the number, weight and Z averages are readily calculated. It would be advantageous if other parameters could be similarly derived. Since viscosity and G.P.C. are both measures of hydrodynamic volume, it would be expedient to obtain intrinsics by merely injecting solutions into the G.P.C. system.[1] In this paper I present data to examine the experimental feasibility of using G.P.C. as a means to obtain intrinsic viscosity and its application to absolute molecular weight determinations and branched polymers.

EXPERIMENTAL

Materials

The fractions and polymers selected for this study include polystyrene fractions (Pressure Chemical Co.), two polystyrene polymers (National Bureau of Standards) NBS 705, which is of narrow molecular weight distribution, and NBS 706, a polymer of wide molecular weight distribution, two whole polyethylene standards, linear NBS 1475 and branched NBS 1476, two polycarbonate standards (General Electric) and two polycarbonate samples received in our laboratory for molecular weight analyses.

Viscometry

All the intrinsic viscosities were measured using a Cannon Ubbelohde semi-micro viscometer. The intrinsics of the two polyethylenes (1475[2] and 1476[3]) were measured at the National Bureau of Standards.

Gel Permeation Chromatography

Two G.P.C. instruments were employed in this study. The Model 200 (Waters Associates) was used for the polyethylenes. The solvent is trichlorobenzene at 130°. The column packing is Styragel whose pore sizes are 10^6, 7.0×10^4-5×10^5, 1.5×10^4-5×10^3, and 2×10^3-7×10^2. The Perkin Elmer High Speed G.P.C. was used for the polymers which are soluble at room temperature. This instrument contains 4 columns packed with Vit-X(a surface treated porous glass) of sizes 10^6, 10^5, 10^4 and 10^3 Å.

A PDP8 Lab 8/E minicomputer (Digital Equipment Corporation) is interfaced with both G.P.C.'s and is programmed for data acquisition and data reduction.

RESULTS AND DISCUSSION

A. G.P.C. Intrinsic Viscosities

With conventional G.P.C. data anlysis, the number, weight and Z averages are calculated from the relationship between the amount of polymer $[W_i]$ and the elution volume corresponding to a particular molecular weight $[M_i]$. These molecular weights have been predetermined by a calibration with known standards.

The analysis can be extended to obtain intrinsic viscosities and $\overline{M}_V$ average molecular weights if the Mark Houwink coefficients (K and a) are known for the polymer-solvent system. The derivation is shown in Appendix A.

The final equation is:

$$[\eta] = \frac{K\Sigma W_i M_i^{\ a}}{\Sigma W_i}$$

The existing computer program for processing the data has been adapted to include the numerical solution of this viscosity equation.

Intrinsic viscosities from the G.P.C. can now be calculated in addition to the usual molecular weights.

The polystyrene fractions were eluted on the Perkin Elmer High Speed G.P.C., with tetrahydrofuran as solvent. The calibration curve was derived from the data using a linear least squares fit (Log M_w vs. V_e). The intrinsic viscosity for each fraction was also calculated. The Mark Houwink coefficients are those reported by Lyngaae and Jorgensen[4] and used in previous work.[5] This equation has been found to be sufficiently accurate for our use.

$$[\eta]^{25}_{THF} = 1.14\ 10^{-4}\overline{M}_v^{.72}$$

The two polystyrene polymer standards were similarly run and analyzed.

The G.P.C. intrinsics obtained are compared to the experimentally measured intrinsics in Table I and Figure 1. The calculated values are in good agreement with the values obtained from the viscometric measurement. The narrow molecular weight polymer shows a difference of approximately 2% between the two and for the broad sample, approximately 7%. This sample was run in triplicate.

The same experimental procedure was followed for linear polyethylene, NBS 1475. The K and a values used[6] for the G.P.C. analysis are those reported by Gilding.

$$[\eta]^{135}_{TCB} = 5.1\times10^{-4}\overline{M}_v^{.76}$$

The G.P.C. intrinsic was then compared to the value published by the National Bureau of Standards.[2]

The agreement for the linear polyethylene sample between the two methods is excellent. The data are shown in Table II. The difficulty of measuring viscosity at 135°C is compounded by filtering and pipetting at this temperature, as well as the care and time needed to bring the volume to the exact mark at the exact temperature. This sample of NBS 1475 linear polyethylene was eluted five times. The precision is excellent and agreement is good between measured and G.P.C. intrinsics. The detector had been changed to an infrared one[7] and the improvement in baseline stability is a major factor in improved chromatograms.

Intrinsic viscosities can now become a part of the routine G.P.C. analysis in addition to the other molecular weight averages if an absolute calibration and adequate Mark Houwink coefficients are

TABLE I

Sample	Polydispersity	True $[\eta]\frac{dl}{gm}$	G.P.C. $[\eta]$
	Polystyrene Fractions		
2×10^6	1.30	3.92	4.16
6.7×10^5	1.15	1.79	1.73
4.11×10^5	1.10	1.28	1.23
1.60×10^5	1.06	0.62	0.68
5.1×10^4	1.06	0.28	0.24
1.98×10^4	1.06	0.138	0.136
	Polystyrene Polymers		
NBS #705 (179,000)	1.07	0.740	0.728
NBS #706 (258,000)	2.10	0.931	[1.01, 0.967, 1.03] 1.00 av.

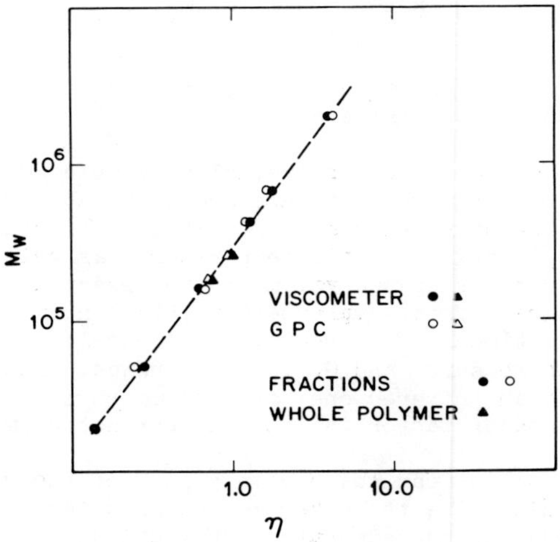

Figure 1. Polystyrene viscosity vs M_W.

TABLE II

Linear Polyethylene

Sample	Polydispersity	True [η] $\frac{dl}{gm}$	G.P.C. [η]
NBS 1475	2.90	1.01*	0.970 1.010 1.012 1.016 1.014

*Measured at the National Bureau of Standards

Branched Polyethylene

Sample	λ	$\overline{M}n \times 10^{-4}$	$\overline{M}w \times 10^{-4}$	True Intrinsic dl/gm	G.P.C. Intrinsic
NBS 1476*	5×10^{-5}	2.12	7.75	.9024	.9024
NBS 1476**	$6\text{-}8 \times 10^{-5}$	1.95	9.09	.9024	.895

* GPC - BTL

** GPC - NBS

available. By an injection technique, once the polymer solution goes onto the G.P.C. columns, no further work is necessary thus eliminating the interaction time spent on normal viscosity measurements. With the use of a computer the data are reduced to include this additional parameter.

B. Applications of G.P.C. Intrinsic Viscosity

The facility of obtaining a G.P.C. intrinsic viscosity expands the G.P.C. analyses in two ways:

(1) Absolute Calibration Curves

Until now, polycarbonate molecular weights have been reported by us in terms of the polystyrene calibration as no wide molecular weight range standards are readily available. It would be more meaningful to the scientist if he could know these absolute values when he is trying to relate other physical properties to molecular weight.

To obtain this absolute calibration, the viscometric intrinsic measurement of one well characterized standard (GE Lexan 62) was used to match the G.P.C. intrinsic viscosity. The equation used[8] was

$$[\eta]^{25°C}_{CHCl_3} = 1.2\times10^{-4}\bar{M}_v^{.82} .$$

The existing polystyrene calibration curve was altered until the two intrinsics matched. The method by which the calibration was changed is shown in Appendix B. The assumption was made that the slope of the polycarbonate calibration is the same as the polystyrene.

The data are shown in Table III. With only one adjustment of the calibration curve, the G.P.C. numbers were obtained in our laboratory for the GE samples Lexan ND-62 and Lexan NE-44, and they were found to be in excellent agreement with the GE reported values. The number average is the most sensitive to small changes in the low molecular weight region of the distribution curve[9], and the significant difference between the values obtained at the two laboratories is not surprising. The other averages agree well.

Two polycarbonate samples received in our laboratory for molecular weight determinations were used for this study. The second half of Table III shows the data. The viscosities of samples designated as A and B were measured by the viscometer. Using the polycarbonate calibration, the G.P.C. intrinsics were found to agree closely with the viscometric intrinsics, confirming the correct values for the molecular weight averages reported for these polycarbonates.

The procedure now to obtain the molecular weights for polycarbonate is to elute the standard whose viscosity has been determined once and for all times, and alter the existing calibration until the viscometric and G.P.C. intrinsics match.

TABLE III

POLYCARBONATE STANDARDS

	G. E. Reported Values				G.P.C. Match		
Sample	$\overline{M}n \times 10^{-4}$	$\overline{M}w \times 10^{-4}$	$\overline{M}z \times 10^{-4}$		$\overline{M}n \times 10^{-4}$	$\overline{M}w \times 10^{-4}$	$\overline{M}z \times 10^{-4}$
Lexan ND-62	1.23	3.16	5.36		1.56	3.18	5.73
Lexan NE-44	1.06	2.76	4.71		1.27	2.71	4.94

POLYCARBONATE SAMPLES

Sample	[η] Measured	[η] G.P.C.		$\overline{M}n \times 10^{-4}$	$\overline{M}w \times 10^{-4}$	$\overline{M}z \times 10^{-4}$
A	$.51_1$	$.51_6$		1.07	2.89	5.50
B	$.54_4$	$.54_4$		1.23	3.08	5.96

A detailed study[10] was made of the melt flow rate of polycarbonate polymers as a function of its weight average molecular weight obtained from G.P.C. With the absolute polycarbonate calibration the results from the G.P.C. runs were plotted (Fig. 2). The slope of this plot calculated to be -3.7 which agreed well with published data.[13] We plan to extend this study to the molecular weight range below the critical value for entanglement.

(2) Branching

The approach to the study of whole branched polymers has been attempted by matching viscometric and G.P.C. intrinsics. The computer program was again modified to include the long chain branching coefficient, λ. Corrections for column spreading and short chain branching were also included. The well characterized branched polyethylene (NBS 1476) was used for the analyses. An iterative procedure was then used in adjusting λ (the long chain branching coefficient) until the G.P.C. intrinsic matched the NBS value, following the procedure of Drott and Mendelsohn.[11] The λ obtained was 5×10^{-5} versus the $6\text{-}8 \times 10^{-5}$ reported by NBS.[12] The molecular weights are reported in Table II.

The study of branched polymers is being investigated further.

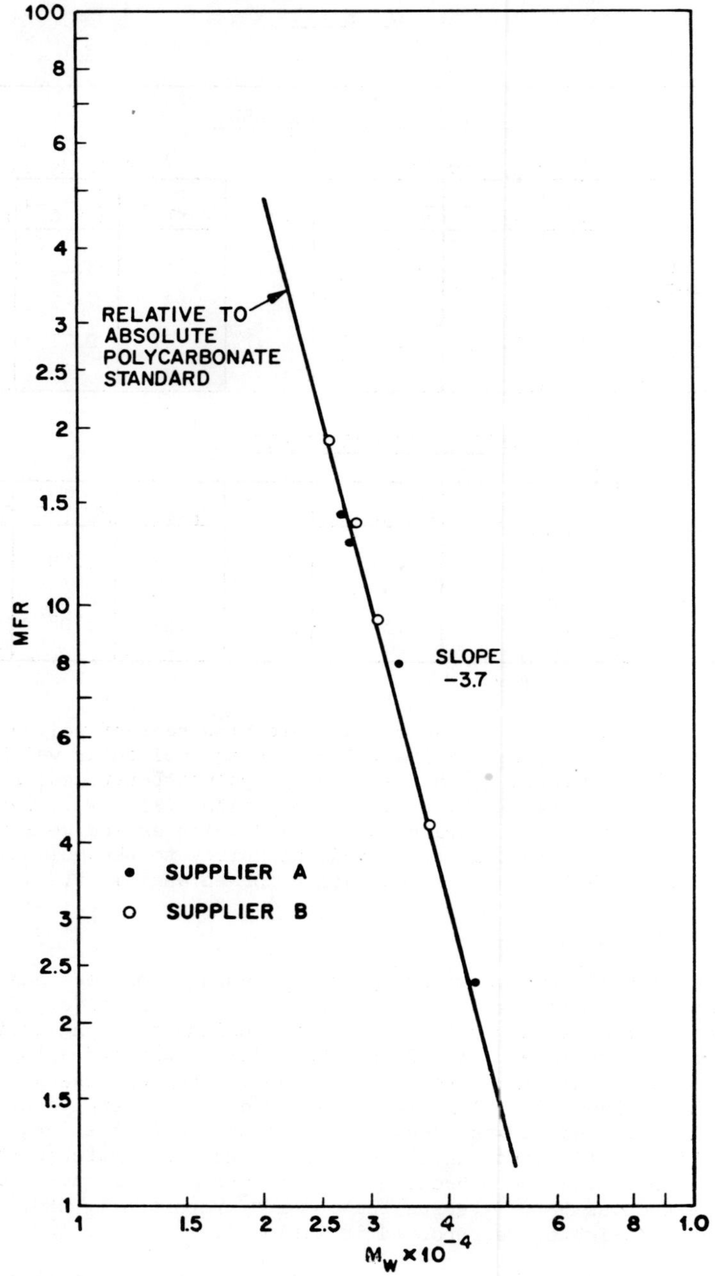

Figure 2. Graph of MFR vs M_w of polycarbonate.

CONCLUSIONS

1. Intrinsic viscosities by G.P.C. can be done on a routine basis for a polymer when an absolute calibration and reliable Mark Houwink coefficients are available.

2. Absolute calibrations can be obtained from one well characterized standard whose intrinsic viscosity is known, again having reliable Mark Houwink coefficients.

3. An approximation of the molecular weights and long chain branching coefficient (λ) of a branched polymer can be made using an iterative computer process until the measured and G.P.C. intrinsic viscosities match. The accuracy of the branching formulae and the assumption that λ is independent of molecular weight are critical to this analysis.

APPENDIX A

Equation 1 is from Reference 1.

$$1.\quad \overline{M}_v = \left[\frac{\Sigma N_i M_i^{(1+a)}}{\Sigma N_i M_i}\right]^{1/a}$$

taking a power of both sides and multiplying them by K.

$$2.\quad K\overline{M}_v^{\,a} = K\left[\frac{\Sigma N_i M_i M_i^{\,a}}{\Sigma N_i M_i}\right]$$

since

$$N_i M_i \equiv W_i \text{ and } K\overline{M}_v^{\,a} \equiv \eta$$

$$3.\quad \eta = K\left[\frac{\Sigma W_i M_i^{\,a}}{\Sigma W_i}\right]$$

APPENDIX B

Calibration Correction

1. Existing polystyrene calibration is used

$$M = A_{PS} + B_{PS} (V_e)$$

2. Mark Houwink equation is

$$[\eta] = KM_v^{\,a}$$

3. Adjust A by

$$A = A + \frac{1}{a} \ln \frac{[\eta] \text{ measured PC}}{[\eta] \text{ G.P.C. PS}}$$

4. B is assumed to remain unchanged.

REFERENCES

1. J. Cazes and J. D. Carter, Indus. Res. 17(7), pp 53-58 (July 1975)

2. C. A. J. Hoeve, H. L. Wagner, and P. H. Verdier, Journal of Research of the National Bureau of Stds. - A. Physics and Chemistry, Vol. 76A, No. 2, March-April, pp 137(1972)

3. C. A. J. Hoeve, H. L. Wagner, J. E. Brown, R. G. Christensen, and J. R. Maurey, Certificate, National Bureau of Stds., 1969

4. J. Lyngaae and J. Jorgenson, J. Chromatog. Sci., 9, 331 (1971)

5. E. P. Otocka and M. Y. Hellman, J. Polym. Sci., B, 12(6), pp 331-9(1974)

6. E. P. Otocka, R. J. Roe, M. Y. Hellman and P. M. Muglia, Macromolecules, Vol. 4, July-Aug, pp 507(1971)

7. J. H. Ross, Jr. and R. L. Shank, Advances in Chemistry Series, No. 125, 1973

8. J. Brandrup, E. H. Immergut, Polymer Handbook Second Edition IV-25, Wiley Interscience, (1975)

9. M. R. Ambler, and R. D. Mate, J. Polym. Sci. A-1, 10(9) pp 2677-2689 (1972)

10. J. Ryan, Society of Plastics Engineers 34th Annual Tech. Conf., pg 205(1976)

11. E. E. Drott and R. A. Mendelsohn, J. of Polym. Sci. Part A-2, Vol 8 1373-1385(1970)

12. H. L. Wagner and F. L. McCracken, Polymer Prepr, A.C.S., Div. Polym. Chem. 16(2), pp 35-40(1975)

13. J. W. Shea, E. D. Wilson, and R. R. Cammons, 33rd SPE Annual Tech. Conf., pg. 614(1975)

USE OF HEXAFLUORO-2-PROPANOL AS A GPC SOLVENT

E. E. Drott

Monsanto Textiles Company
Pensacola, Florida

ABSTRACT

Hexafluoro-2-propanol (HFIP) has proven to be a useful gel permeation chromatography (GPC) solvent for a number of polymers which have hydrogen bonding sites. Its UV transmission, refractive index, viscosity and solubility properties permit its use with a variety of GPC detectors and columns. Examples of polyester, polyamide and polymethylmethacrylate characterization are given.

* * * * * * * * * * *

In gel permeation chromatography (GPC) separations, several basic requirements for a solvent must be met. The solvent should: 1) dissolve the polymer being characterized without degradation during dissolution or sample fractionation; 2) not significantly reduce column efficiency; 3) not lead to adsorption of sample on column and packing surfaces; 4) have a relatively low viscosity; and 5) have physical properties that permit sample detection as it elutes from the column.

Usually several solvents can be found that fulfill these requirements for a given polymer. The final choice of solvent is an optimization of the required properties against available equipment, types of polymers to be characterized and economics of the analysis (manpower and solvent cost).

Solubility and viscosity properties have limited the number of solvents that have been successfully used for GPC separation of nylons, polyesters and other polar polymers (1,2,3). Elevated temperatures are required in many cases to dissolve the polymer or reduce solvent viscosity. Mixed solvents have been used which permit operation at lower temperatures (4,5,6). However, the latter introduces further problems in reproducing solvent ratios which may affect polymer hydrodynamic volume and detector response.

Hexafluoro-2-propanol (HFIP) has been found to be an excellent GPC solvent for nylon, polyesters and other polar polymers. Polymer dissolution and GPC separations are made at room temperature. Sample elution can be detected by either UV or differential refractometer detectors.

Multimodal GPC curves have been observed in polymer systems where polyelectrolyte effects can occur (7). In several of the HFIP polymer systems investigated, multimodal data were obtained. As in the case of dimethylformamide and other polar solvents (7), addition of a salt to the HFIP suppressed the polyelectrolyte effect and normal distributions were obtained.

Although only results for nylon 6,6, poly(ethylene terephthalate) (PET) and polymethyl methacrylate (PMMA) are discussed in this paper, several other commercial and experimental polymer types have been successfully characterized by GPC using HFIP as the solvent.

PHYSICAL AND SOLVENT PROPERTIES

HFIP is a volatile, polar liquid having a high density and low viscosity. It has a high degree of thermal stability and is noncorrosive. It is transparent to UV light (< 2000Å). Physical properties are listed in Table I (8).

HFIP exhibits strong hydrogen bonding and will associate with and dissolve most molecules with receptive sites. It is soluble in

TABLE I

PHYSICAL PROPERTIES OF HFIP

Melting Point	-3.4°C (26°F)
Boiling Point (760 mm)	58.2°C (136.8°F)
Specific Gravity (25°C)	1.59 g/cc (13.3 lb/gal)
Refractive Index (n_d^{25})	1.2752
Surface Tension (25°C)	16.3 dynes/cm
Viscosity @ 25°C	1.021 cs
@ 38°C	0.670 cs
Flash Point	Nonflammable
Fire Point	Nonflammable

many organic solvents, but is immiscible with long chain alkanes and some aromatic compounds. Hence, consideration should be given to compatability of HFIP with other compounds when changing solvents in the GPC columns.

HFIP dissolves a wide range of polymers. The solubilities of a number of these in HFIP are given in Table II.

TABLE II

SOLUBILITY OF POLYMERS IN HFIP

Polymer	Solubility*, g/100 gm @ 25°C
Polyamides	25 - 27
Polyesters	15 - 20
Polyacrylonitriles	7 - 8
Polyacetals	> 5 - 7
Polymethylmethacrylates	25 gel
Hydrolyzed Polyvinylesters	15 - 20

* Reference 8.

EXPERIMENTAL

Two GPC systems were used in this study. Data for polymers run in tetrahydrofuran (THF) were obtained on a Waters Model 301 GPC equipped with a differential refractometer. A system consisting of a LDC 5000 psi pump, Waters UK6 injection valve, Schoeffel Model 770 UV detector and Waters Model R401 differential refractometer was used for HFIP investigations.

Because of the high cost of HFIP (~ $550/gallon), effluent from the detectors was fed directly to a distillation pot. A simple distillation was used to recover the HFIP which was then pumped to the GPC solvent reservoir. Between 500 and 1000 samples per gallon of HFIP can be characterized using this technique.

GPC separations in HFIP were made using a set of μ-Styragel (9) columns (10^3 - 10^4 - 10^5 - 10^6 Å). Pressure drop across the column set was approximately 3500 psi at a flow of 2 ml/min. at 25°C.

Although HFIP is a poor solvent for the polystyrene-packing (does not swell gel), column efficiency (1400 plates/column set) has remained stable for two years.

GPC data were also obtained with a set of porous glass columns manufactured by DuPont (10). Pressure drop across the three column set (100 - 500 - 1000 Å) with HFIP was approximately 1000 psi at 1 ml/min. at 25°C. A plate count of 9000 plates was calculated for the column set.

Elution time (injection to injection) for both sets of columns was thirty minutes. Data reduction was carried out with an IBM 1800 data acquisition system. GPC peak heights were digitized every six seconds and variation in elution volume times corrected by means of a small molecule internal standard (dimethyl terephthalate).

POLYSTYRENE GEL VS. POROUS GLASS COLUMNS

As noted in the preceding section, the glass columns have a higher efficiency than the polystyrene gel columns as calculated from the dispersion of the dimethyl terephthalate internal standard. The high molecular weight exclusion limit for the glass columns is lower than that for the gel columns. However, from a practical viewpoint, both column sets have acceptable resolution and efficiency for useful separations with HFIP. Raw GPC curves for a PET sample are shown in Figures 1 and 2 and illustrate this point.

POLYMER AND COLUMN STABILITY

It has been reported that PET is degraded when it is dissolved in HFIP (6). Our data and other information (8) indicate that HFIP dissolves nylon, PET and other polymers without any degradation. PET samples which had been dissolved and recovered from HFIP had the same intrinsic viscosity in phenol-tetrachloroethane as the starting material. In other experiments, HFIP - polymer solutions were aged for several weeks with no apparent change in color, intrinsic viscosity or GPC data. Reproducibility of nylon data are illustrated in Figure 3 and in Table III.

Fluorinated solvents (trifluoro-ethanol, HFIP, etc.) are not compatible with the polystyrene gels used for GPC packings, i.e., the gels tend to shrink rather than swell in the solvent. This shrinkage can decrease or destroy column efficiency, especially in the older Styragel columns (37µ, 3/8 in. x 4 ft.)(3). In the case of the µ-Styragel columns, the smaller gel particles (10µ) yield a higher packing regularity which apparently reduces the formation of channels and voids when HFIP is used and the columns maintain their efficiency.

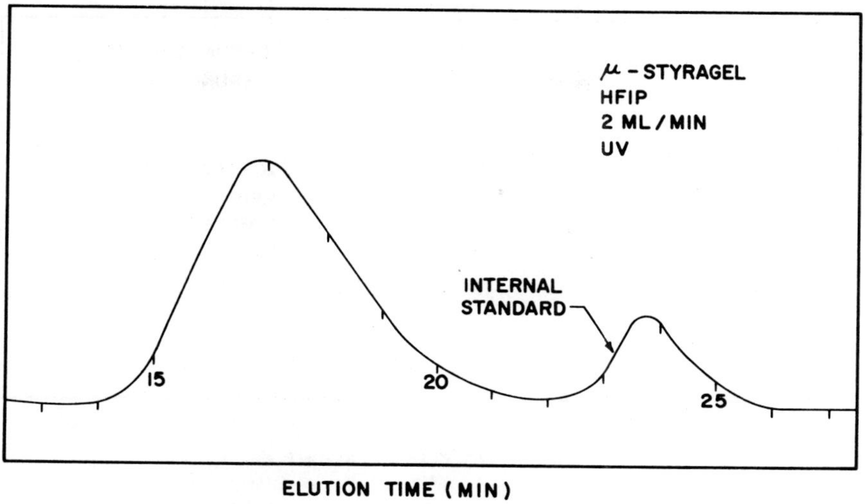

Figure 1 - Raw GPC curve for PET obtained with μ-Styragel columns.

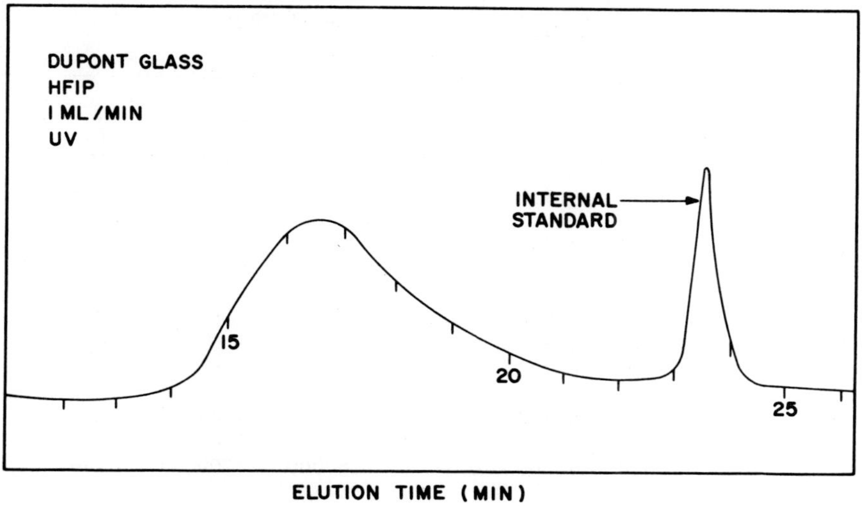

Figure 2 - Raw GPC curve for PET obtained with DuPont glass columns.

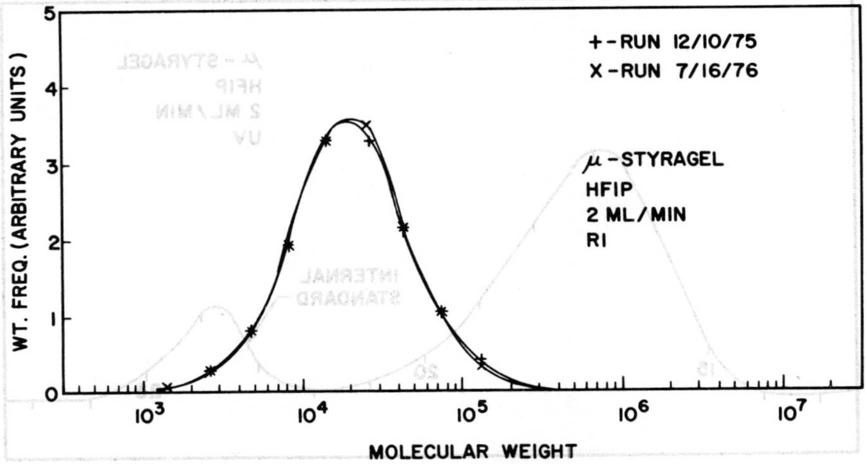

Figure 3 - Reproducibility of nylon 6,6 GPC data.

TABLE III

SUMMARY OF MWD AND VISCOSITY DATA

Sample	GPC (HFIP)			Non-GPC		
	M_n	M_w	[η]	M_n	M_w	[η]
Nylon 6,6 (12/10/75)	13100	31300	1.12	--	--	--
Nylon 6,6 (7/16/76)	12900	30700	1.11	--	--	--
PET No. 1	1540	3600	0.11	--	--	0.11
PET No. 2	18500	42300	0.65	--	--	0.64
PET No. 3	29700	86300	1.07	--	--	1.00
PMMA 6038	20200	56200	--	19400	49200	--
PMMA 6036	43600	113000	--	48600	114000	--
PMMA 6041	121000	253000	--	160000	267000	--
PMMA 6038	24300*	45400*	--			
PMMA 6036	49500*	108000*	--			
PMMA 6041	144000*	273000*	--			

*THF.

POLYELECTROLYTE EFFECTS

Several polymer types exhibit multimodal GPC curves in "wet" HFIP in the absence of an ionizing salt as illustrated in Figure 4 for nylon 6,6. Sodium trifluoroacetate (NATFAT) was found to suppress the polyelectrolyte effects in all of the HFIP-polymer systems that have been studied. Since the same raw GPC data were obtained with 0.01M and 0.08M solutions, the lower concentration (~0.8 gm NATFAT/kgm HFIP) was selected for routine analysis.

The lack of a salt concentration dependency on HFIP solution properties suggests that the hygroscopic NATFAT may be immobilizing the water in the HFIP and preventing it from acting as an ionizing solvent. It is also possible that the 0.01M concentration was higher than that necessary to neutralize polymer-solvent interactions. Other salts were not investigated since NATFAT proved successful in the GPC separations.

CALIBRATION

Direct calibration for nylon 6,6 and PET whole polymers can be easily carried out since the standard polymers have relatively narrower MWD's (~2) and Mark-Houwink constants are well defined. The calibration curves for nylon and PET shown in Figure 5 were obtained by trial and error adjustment of the coefficients of a cubic polynomial until intrinsic viscosities calculated from GPC data for the samples agreed with experimentally measured values. Typical GPC data for three PET samples are given in Figure 6 and Table III.

Since polystyrene is insoluble in HFIP, universal calibration techniques based on elution volumes of narrow MWD polystyrene standards and polymer intrinsic viscosity cannot be used. However, one can use the method presented by Provder (11) where a polymer that is soluble in THF and HFIP is used as a standard. Polymethylmethacrylate is soluble in both solvents. GPC data for PMMA in THF and HFIP are shown in Figures 7 and 8 and Table III.

TOXICITY AND HANDLING PRECAUTIONS

One of the major concerns with use of compounds that have not had wide spread use or been subjected to extensive toxicity testing is their potential hazards. A limited amount of data have been reported for HFIP (8). The principle danger is in skin-contact which results in eye-damage and chemical burns.

HFIP is presently being used with proper handling and ventilation in a number of applications with no observed toxicity problems. Its use should be confined to a laboratory hood or where exhaust ventilation ducts prevent escape of vapor into the air that is breathed by personnel. Adequate protective equipment to prevent any eye or skin contact should be used.

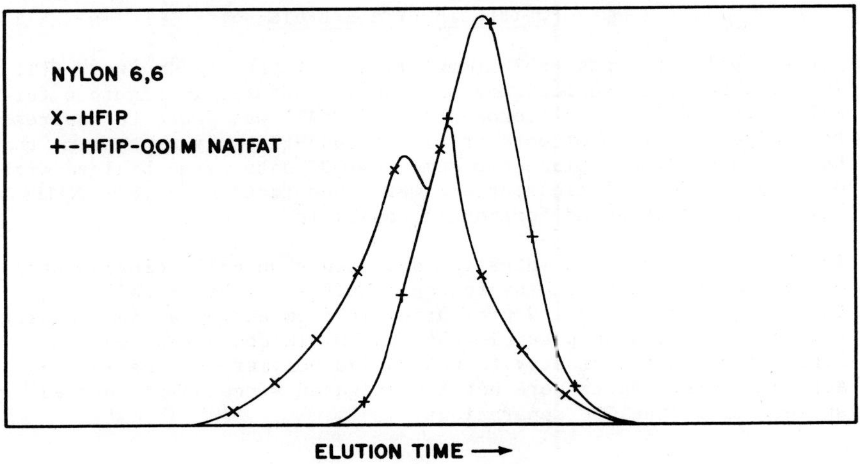

Figure 4 - Suppression of polyelectrolyte effect with NATFAT.

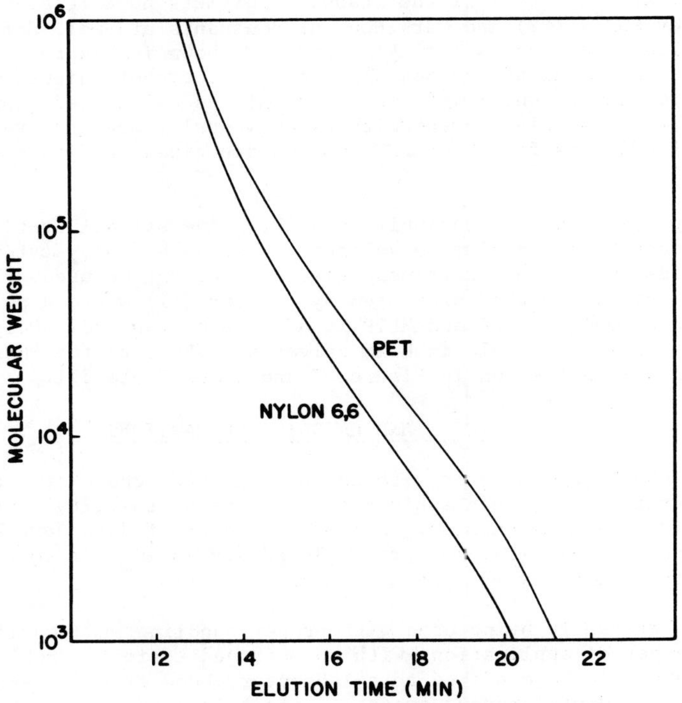

Figure 5 - Calibration curves for μ-Styragel columns.

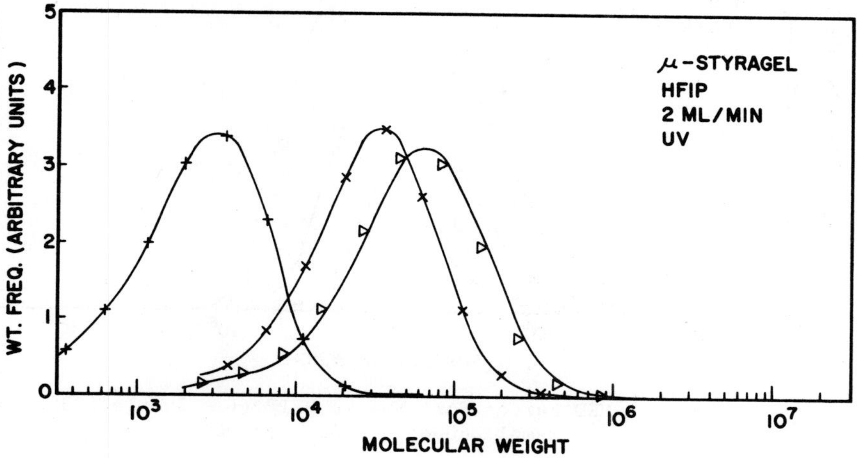

Figure 6 - MWD curves for PET (+ - 1, x - Sample 2, Δ - Sample 3).

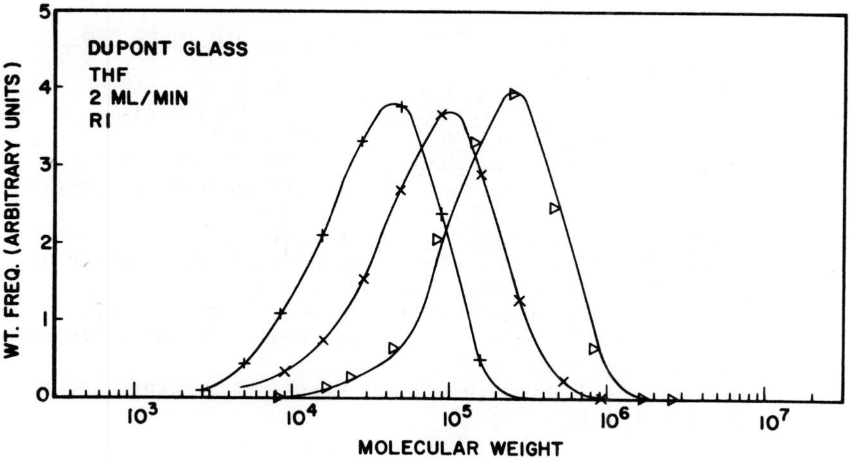

Figure 7 - MWD curves for PMMA in THF (+ - Sample 6038, x - Sample 6036, Δ - Sample 6041).

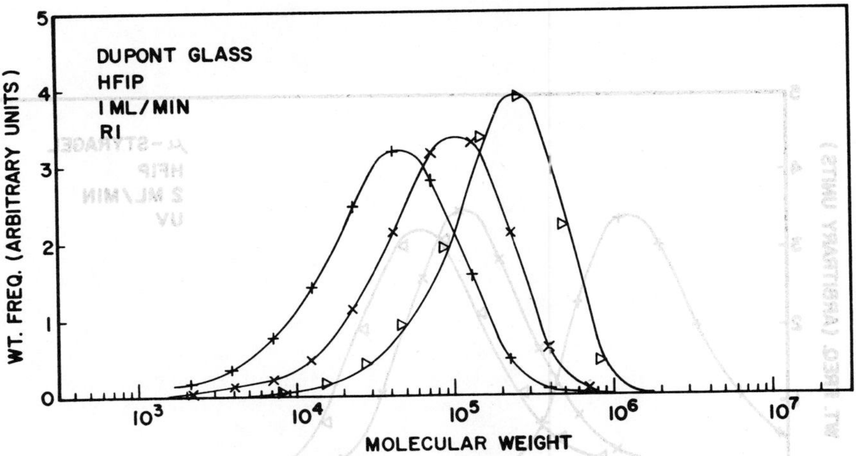

Figure 8 - MWD curves for PMMA in HFIP (+ - Sample 6038, x - Sample 6036, Δ - Sample 6041).

REFERENCES

1. J.R. Overton, J. Rash, L.D. Moore, Proc., 6th Inst. Sem. on GPC, Miami, Fla. (1968).
2. R. Panaris, G. Pallas, J. Polym. Sci., B, 8, 441 (1970).
3. M. A. Dudley, J. Appl. Polym. Sci., 16: 493 (1972).
4. P. S. Ede, J. Chromatogr. Sci., 9: 275 (1971).
5. Y. Ishida, K. Kawai, Shimadzu Hyoron, 29112: 89 (1972).
6. E. E. Paschke, J. G. Bermann, B. A. Bidlingmeyer, Polymer Preprints, 17: 440 (1976).
7. C. Y. Cha, J. Polym. Sci., B, 7, 343 (1969).
8. "Product Bulleting - Hexafluoroisopropanol", E.I. DuPont de Nemours & Co., Wilmington, Delaware (1968).
9. Waters Associates, Inc., Milford, Massachusetts 01757.
10. E. I. DuPont de Nemours and Co., Instrument Products Division, Wilmington, Delaware 19898.
11. T. Provder, J. C. Woodbrey, J. H. Clark, E. E. Drott, Advan. Chem. Ser. No. 125, 117 (1973).

PREPARATIVE FRACTIONATION OF LIQUID CRYSTALS

T. C. Huard

Waters Associates, Inc.
Milford, Massachusetts

Liquid crystals, although they are solid substances, contain molecules that are not oriented in one fixed position; rather the molecules can vary their orientation in response to changes in the environment, such as temperature, electrical fields, light, etc. In other words, the molecular species behave as though they are in a liquid state, although they actually comprise a solid. Because of these unique properties, liquid crystals are particularly useful in applications where the optical changes that accompany the changes in molecular orientation yield special effects, i.e., color changes resulting from temperature variations, changes in reflectivity resulting from modification of an electrical field, etc. The familiar liquid crystal wristwatch and the colorful digital thermometers are good examples of products that make use of these properties.

The unusual properties of liquid crystals can lead to a variety of novel products. Imagine, a car paint that changes color with the day to day temperature variations, or even the seasonal variations. Think of the convenience of using liquid crystal tape or paint to monitor the potential overheating of an apparatus. A simple change in color would signal a dangerous temperature increase. No longer would it be necessary to rely on temperature meters which could malfunction or be misread.

Research has shown that the purity of a liquid crystal determines, in large measure, the quality and lifetime of the product in which it is used. Physical properties such as resistivity, clearing point depression, transition point, etc., are all affected by the purity of the liquid crystals. In most applications, a very small amount of an impurity will give the liquid crystal undesirable characteristics. It is important, therefore, to use materials which possess purities as close to 100% as possible. It is equally important to obtain this high purity level reproducibly. For most applications, 99.5% purity is not suitable. Although not ideal, a purity level of 99.7% to 99.8% is usually the best that can be obtained on a batch-to-batch basis. The liquid crystal industry needs a method to remove these last few tenths of a percent of impurities. This improvement would result in increased product life, reduced reject rate, and increased product quality.

Liquid chromatography (LC), a non-destructive method used to separate complex chemical mixtures on either a small (analytical) or large (preparative) scale (as summarized in Figure 1), is one solution to the problem of obtaining extremely pure liquid crystals. LC is already being used as an important research or production aid in almost every industry.

Of the several different separation mechanisms classed under the general heading of liquid chromatography (LC), the most common are adsorption (both normal- and reverse-phase), gel permeation chromatography (GPC), and ion-exchange. Adsorption chromatography allows the separation of a wide variety of mixtures. Most of the traditional open column liquid chromatography done to date has relied upon adsorption as the mechanism of separation. Figure 2 illustrates the principle of adsorption chromatography. A dissolved sample mixture is carried across the stationary phase by the mobile phase. If there is some attractive force between a component in the mixture and the stationary phase, this component tends to elute from the instrument at a later time. The stronger the attractive force between a component and the stationary phase, the longer the elution time.

	Sample Qty.	Result
Analytical	<1 – 10mg	Data (strip chart) (*e.g.* quality control)
Small-scale Preparative	10 – 200mg	Dilute solutions for further identification (*e.g.* competitive materials analysis)
Preparative	Gram Quantities	Pure materials (*e.g.* product development)

FIGURE 1. The scope of liquid chromatography.

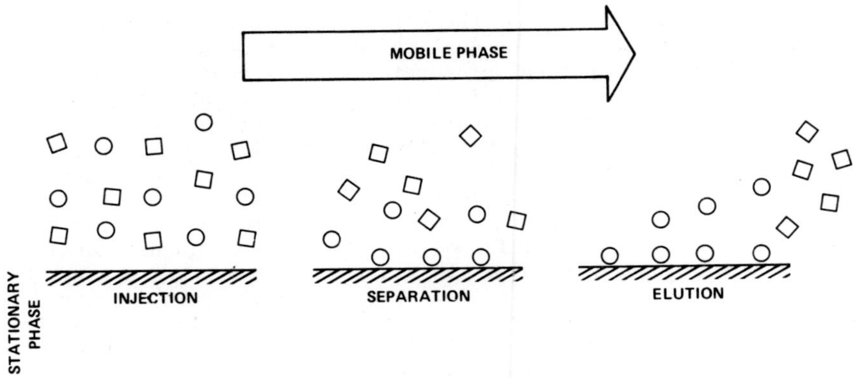

FIGURE 2. The mechanism of a separation in liquid chromatography.

If adsorption chromatography is used in a preparative separation, one can expect results similar to those obtained in the following liquid crystal separation. The general approach to such a preparative separation is outlined in Figure 3. The following discussion will show how this approach is used to separate a liquid crystal mixture.

DEFINITION OF THE PROBLEM - STEP 1 IN THE DEVELOPMENT OF A PREPARATIVE SEPARATION

A commercially available mixture of liquid crystals (E7), has the composition shown in Figure 4. Note that most (77%) of the mixture consists of components A and B and that the chemical difference between these two compounds is slight. The procedure used to apply the mixture to electronic display modules contaminates a large percentage of the expensive liquid crystal mixture. Because contamination greatly impedes the performance of liquid crystals, industry has had to discard thousands of dollars worth of these compounds every year. Recently, Motorola decided that large scale purification of liquid crystals might be profitable and that preparative scale liquid chromatography might provide a solution to this frustrating problem. Motorola wanted to obtain extremely pure fractions of each of the four major components in E7 in multigram quantities.

DEVELOPMENT OF AN ANALYTICAL SEPARATION - STEP 2 IN THE DEVELOPMENT OF A PREPARATIVE SEPARATION

Figure 5 shows an analytical separation of the four major components in E7. The small degree of separation between components A and B reflects the fact that their chemical structures are very similar. The separation was developed on an analytical instrument for a number of reasons. These included the ease of mobile phase modification and subsequent column equilibration, the much smaller mobile phase consumption and sample injection, and the increased resolving power and sensitivity of an analytical system.

An increase in both the polarity of the mobile phase and detection sensitivity revealed the minor and trace impurities that are present in this mixture. Figure 6 compares the response from an injection of solvent (blank) to the response of solvent plus sample. The four major components are contained in the solvent peak. If the responses for all components had been kept on scale, over forty different impurities would have been seen. Identifying any individual component is quite simple. Merely collect the component in question as it elutes from the system. Since liquid chromatography is non-destructive and other interfering components have been removed, the desired component can be identified by any number of ancillary techniques (NMR, UV, IR, etc.).

What is needed to obtain a good preparative separation? An ideal separation would afford narrow, well-separated components which elute from the system in a relatively short time interval.

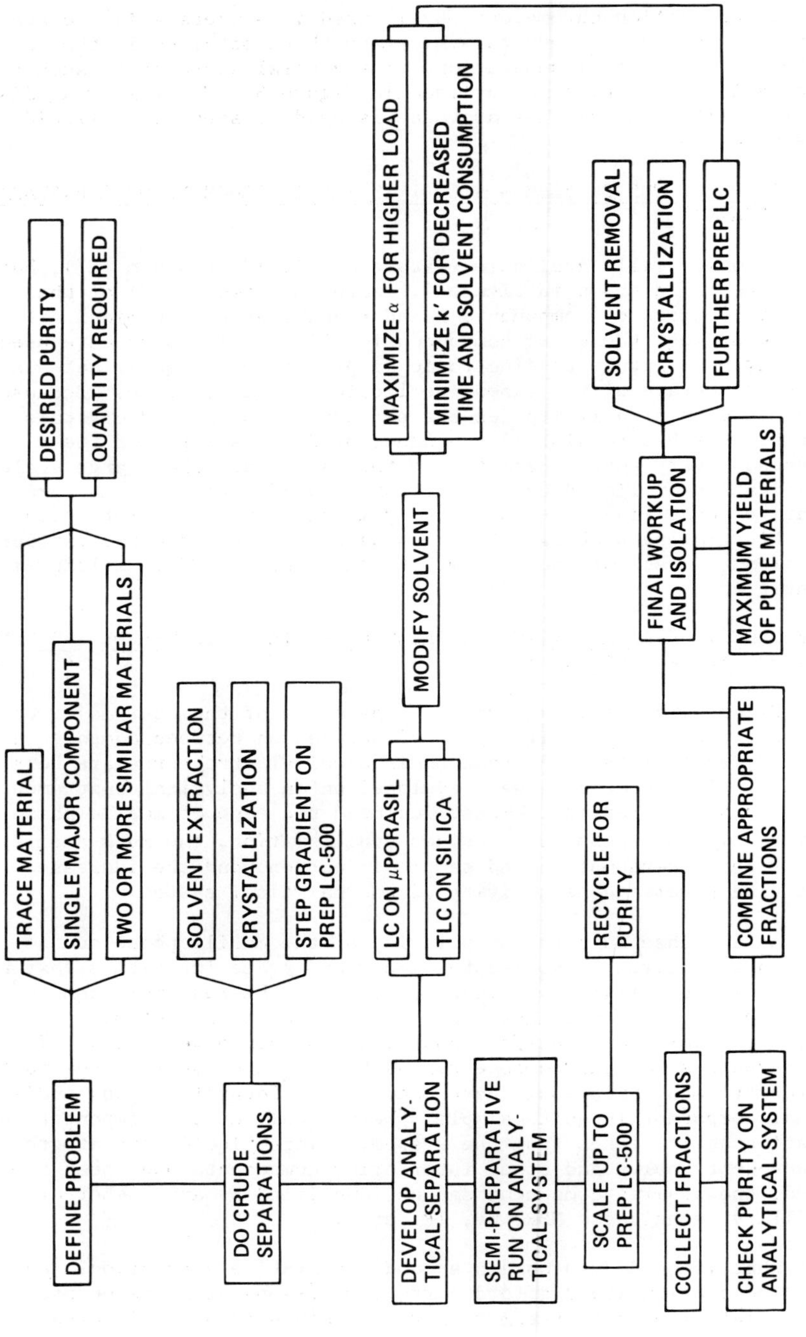

FIGURE 3. How to achieve a preparative separation.

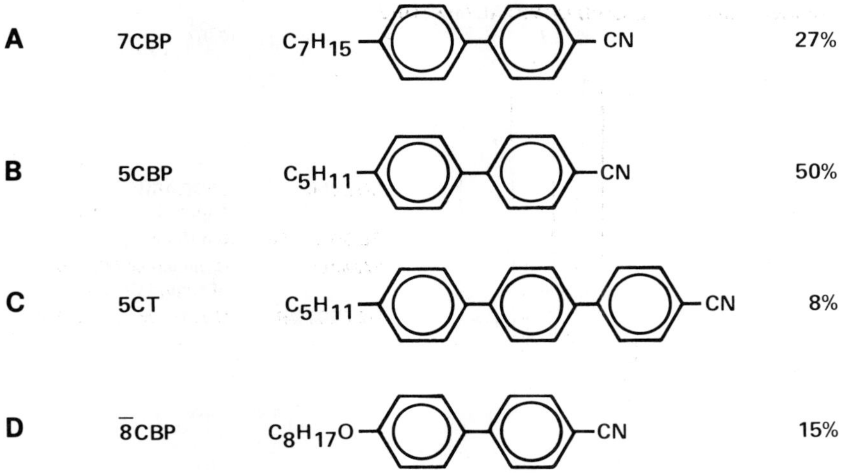

FIGURE 4. Identification of components in E7, a commercially available liquid crystal mixture.

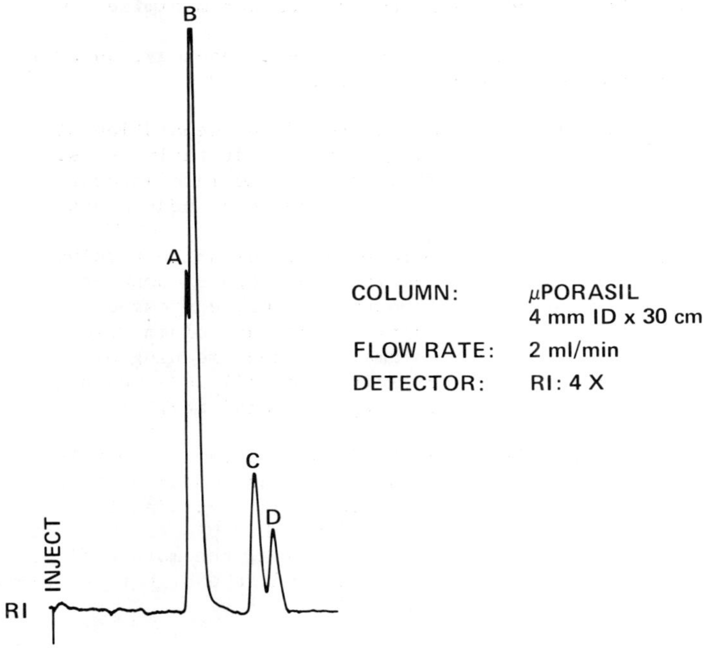

FIGURE 5. An analytical separation of the major components in E7.

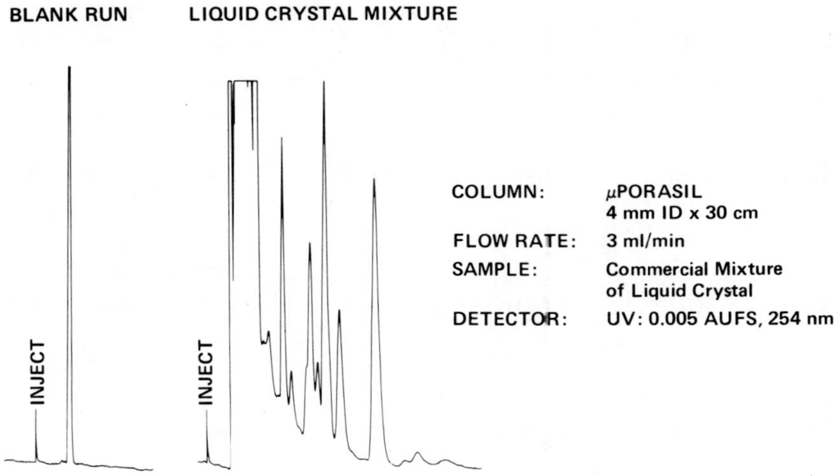

FIGURE 6. Detection of impurities in E7.

Meeting these requirements would enable one to maximize the quantity of material introduced to the system and minimize time.

In order to better understand the technology, an explanation of some of the terminology is in order.

Alpha (α) - A measure of the separation of two components at their peaks. It does not take into account any overlapping of peak areas.

Capacity (k') - A measure of the solvent volume required to elute a component from the column, expressed as multiples of the column void volume. Another meaning of capacity (not k') is a measure of sample loadability.

Efficiency (N) - A standard of column performance relating to the amount of peak spreading that occurs as the separation takes place. The narrower the peak, the more efficient the column and the higher the plate number (N) will be.

Resolution (R) - The total measure of component separation at peak apexes and baselines. Components with R equal to 1.5 are baseline separated.

Figures 7 and 8 show the important role the above terms play in the development of a preparative separation. Sample loadability varies directly with the α value (Figure 7). Figure 8 takes an initial separation and shows the affect of decreasing k' (separation is reduced), increasing k' (separation is increased), increasing N (more efficient column means less spreading and narrower peaks for increased separation), and increasing α (separation is increased).

In preparative liquid chromatography, the k' and α factors can be adjusted by modifying the mobile phase. The column efficiency (N) is greatly dependent upon the column packing technology. Figure 9 shows how radial compression has greatly advanced preparative LC column packing technology. Conventional columns, packed by filling a rigid cylinder with particles that settle along the longitudinal axis, may contain voids and channels. Column to column reproducibility is poor. With radial compression, however, a flexible walled column is compressed by pressurizing the outside of the cartridge wall with gas. This compression packs the column so well that voids and channels are virtually eliminated. Radial compression forms a homogeneous, highly efficient and reproducible chromatographic bed. This feature is standard on the Waters Prep LC/System 500 (Figure 10) used in this work.

MAXIMIZE α FOR HIGHER LOAD & MINIMIZE k' FOR DECREASED TIME & SOLVENT CONSUMPTION - STEP 3 IN THE DEVELOPMENT OF A PREPARATIVE SEPARATION

In order to meet the conditions of high load and minimum time, it was decided to first separate the four components of mixture E7 into two fractions. After this initial separation into mixtures of components A and B and components C and D, each of the two fractions would be reinjected to separate the individual components.

PREPARATIVE SEPARATION AND PURITY CHECK OF COLLECTED FRACTIONS - STEP 4 IN THE DEVELOPMENT OF A PREPARATIVE SEPARATION

The analytical data indicated that a 10 gram injection of E7 was acceptable. Figure 11 is the chromatogram obtained from the preparative separation of E7. The collected fractions are indicated in Figure 11 and the analytical confirmation of purity for one of these fractions, components C and D, is shown in Figure 12.

Components A and B, comprising 77% of the original mixture (E7) have been almost eliminated. Reinjection of this collected mixture and subsequent collection of fractions afforded two pure liquid crystals (Figures 13 and 14). Similar chromatograms will be generated for components A and B as soon as all work has been completed.

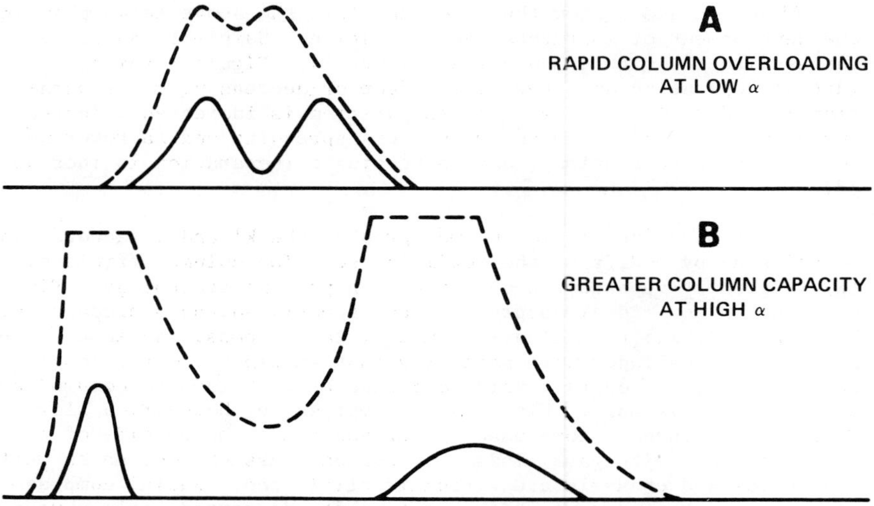

FIGURE 7. The role of α in preparative separations.

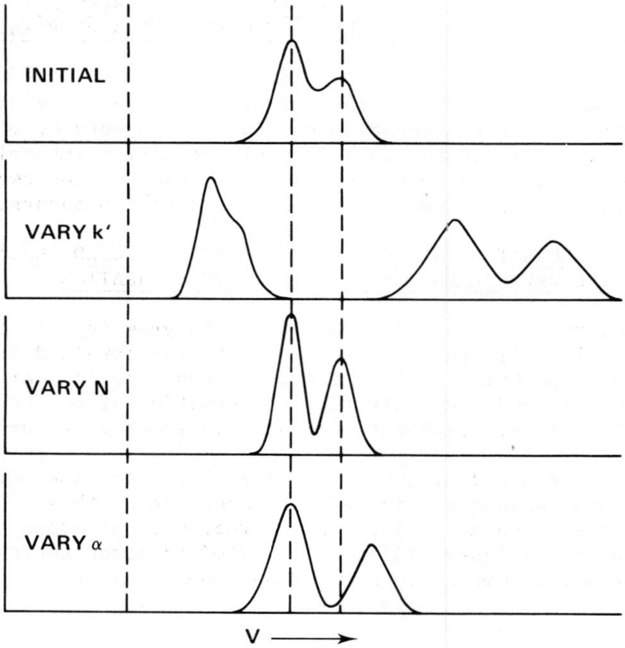

FIGURE 8. Effect of α, k' and N in controlling resolution.

RADIAL COMPRESSION	VERSUS	TRADITIONAL METHODS
Flexible walled cylinder	COLUMN CONSTRUCTION	*Rigid* walled cylinder
Dry powder	METHOD OF FILLING COLUMN	Dry powder or slurry
Particles radially compressed *after* filling in direction *perpendicular* to flow	FORMATION OF BED	Particles settled *during* filling along longitudinal axis, *parallel* to direction of flow
Channels and voids virtually eliminated, cannot form during operation	BED STRUCTURE	Channels and voids, especially near walls may form during packing or operation
Uniformly good and reproducible with minimum operator experience	TYPICAL EFFICIENCY	Poor to good, depending upon skill of operator, choice of method; variable reproducibility

FIGURE 9. A comparison of preparative column packing techniques: radial compression versus traditional.

FIGURE 10. Waters Associates, Inc., Prep LC/System 500.

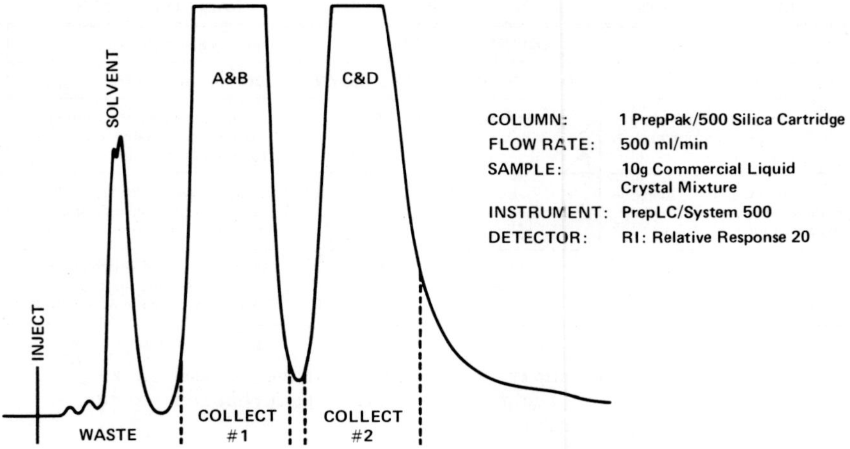

FIGURE 11. A separation of 10 grams of E7 into two fractions.

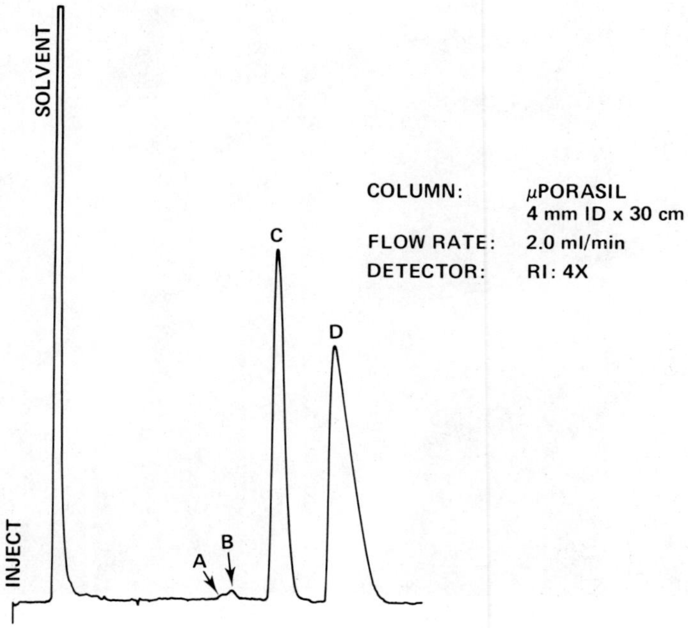

FIGURE 12. Confirmation of purity of C/D fraction on an analytical system.

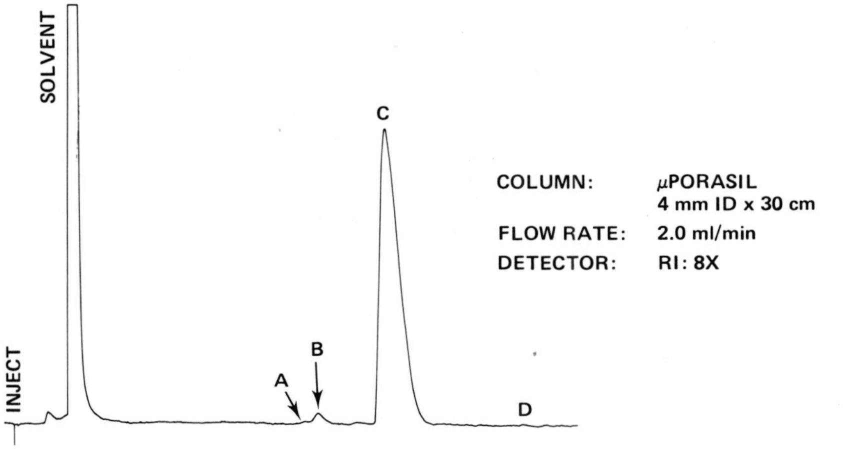

FIGURE 13. Confirmation of purity of component C on an analytical system.

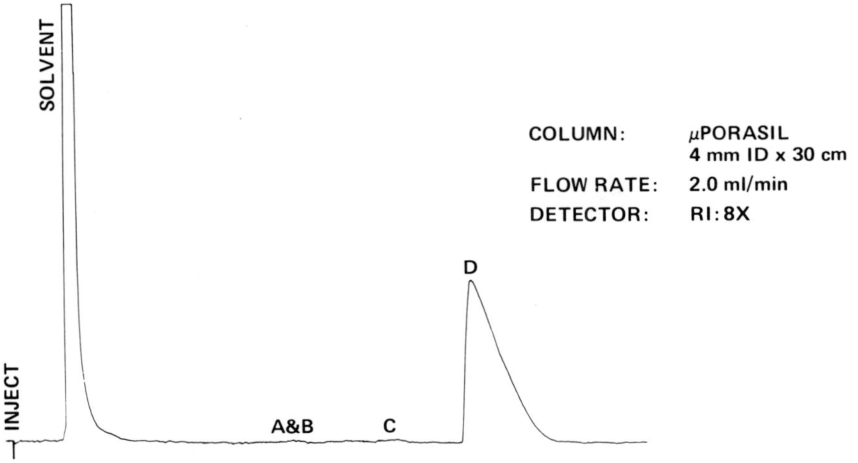

FIGURE 14. Confirmation of purity of component D on an analytical system.

SUMMARY

Obtaining large quantities of pure materials is a problem that has plagued the liquid crystal industry. This work provides the industry with a way to drastically reduce waste, increase product life and quality, and identify impurities present in complex chemical mixtures.

CHROMATOGRAPHIC ANALYSIS OF EPOXY RESINS

D. J. Crabtree
D. B. Hewitt

Aircraft Division
Northrop Corporation
Hawthorne, California

SUMMARY

Chromatographic analyses, using both gel permeation chromatography (GPC) and reversed phase high pressure liquid chromatography (HPLC) were found to be successful methods to analyze epoxy resin formulations which are used in the manufacture of graphite fiber-epoxy resin composite structures. The chromatographic analyses detect presence of epoxy oligomers, curing agents, and impurities. These analyses are important to the aerospace industry because airframe structures which are critical to flight safety are being designed from the graphite fiber-epoxy resin composites.

Epoxy resin formulations based upon N, N, N', N' - tetraglycidyl -4, 4' - methylenedianiline (Ciba-Geigy MY-720) with diaminodiphenyl sulfone curing agent were analyzed by means of GPC and HPLC. A formulation of bisphenol F epoxy (Dow 7818) with a mixture of primary aromatic amine curing agents (Epon Z) was also analyzed.

Both GPC and HPLC separate the resin formulations into the major components (epoxy oligomers and the curing agents) and minor components such as impurities. However HPLC is much more sensitive than GPC in detecting minor components. GPC gives an excellent analysis of the chemical changes which occur in a catalyzed epoxy resin as it ages at low temperature. The development of higher molecular weight components and the disappearance of curing agent are readily detected. The changes which occurred in the Ciba-Geigy MY-720/DDS resin system after up to three years storage were determined by GPC. Chemical changes in the Dow 7818/Epon Z resin system were followed beyond the gel time at room temperature using GPC. Differential thermal analysis was used to establish the presence of unreacted epoxide in the Dow 7818/Epon Z long after the resin had gelled.

Experimental

1. Chromatographic Analyses

Chromatographic analyses were made using Waters Associates Liquid Chromatograph Model ALC/GPC-244. For GPC analyses Styragel columns purchased from Waters were used. Five columns were used in series. The pore sizes of the columns were 100A (2 columns), 500A (2 columns), and 1000A (1 column). The five column set was calibrated using polypropylene glycol standards purchased from Waters Associates. The molecular weights of the standards were 800, 1200, 2000, and 4000. Tetrahydrofuran solvent (ultraviolet grade) purchased from Burdick and Jackson Laboratories Inc., Muskegon, Michigan, was used in all GPC analyses. The instrument was operated at room temperature with a flow rate of 1.0 ml/minute.

Reversed phase high pressure liquid chromatographic analyses were made on the ALC/GPC-244 chromatograph using a Bondapak C_{18}/Corasil 2 mm diameter column purchased from Waters Associates. The instrument was operated at room temperature at a flow rate of 2.0 ml/minute. Distilled water and tetrahydrofuran (Burdick and Jackson) were used as solvents. Analyses were made using the linear gradient on the Model 660 solvent programmer. Program time was 60 minutes with the solvent mixture varying from 10% tetrahydrofuran to 100% tetrahydrofuran over the 60 minutes.

2. Differential Thermal Analyses

Differential thermal analyses were run using the Stone LB-202 differential thermal analyzer (Columbia Scientific Industries) at a heat up rate of 10 C/minute. Samples were analyzed in a flow of argon gas.

INTRODUCTION

The need for light weight, high stiffness, high strength materials for aircraft structures is presenting a severe challenge to epoxy resin technology. Graphite fiber/epoxy resin composites are receiving increasing attention for major structural components for modern military aircraft.

materials. In the future commercial passenger and freight aircraft may be built with the same materials. For approximately twenty-five years some aircraft structures have been built from glass fiber/polyester resin and glass fiber/epoxy composites. However, these structures were not critical to flight safety and their failure did not lead to the loss of an aircraft. The structures now being designed from graphite fiber/epoxy composites, such as wings and fuselage sections, are critical to flight safety. Failure of these composites will lead to loss of aircraft. Therefore all aspects of epoxy resins used in these structures are of critical importance.

It is essential to know the composition of the proprietary resins being purchased in order to determine if the composition changes from batch to batch. Since human life depends upon the performance of the epoxy resin, the quality of the purchased resin must be assessed by the user. Of equal importance are the analyses which should be made on the resin during storage prior to cure to establish what chemical changes occur to the resin and how these changes affect the reliability of the composite structure after it has been manufactured. An analysis of an epoxy resin formulation is complex because in a typical resin formulation one or two base epoxy resins will be used along with a small concentration of a tackifier resin as well as the curing agent and possibly an accelerator. In the ideal analysis the total composition of the resin will be determined with each component being fully identified. As a preliminary step, a "fingerprint" analysis which reveals the presence and relative quantity of all components but does not identify each is acceptable.

Instrumental chromatography, both gel permeation and high pressure liquid chromatography, are ideal methods for these analyses. The analyses are rapid, economical to run, and accurate. This paper presents a discussion of the chromatographic techniques used to perform fingerprint analyses of epoxy resin formulations used in graphite composite manufacture. In addition, results of chromatographic analyses of the chemical changes which occur in the epoxy resins during refrigerated storage at 0 C and during storage at room temperature are presented.

RESULTS AND DISCUSSION

1. "Fingerprint" Analysis of Epoxy Resin Formulations

There are four major epoxy resin formulations being used by the aerospace industry for the manufacture of graphite fiber/epoxy composites. These formulations are based upon the Ciba-Geigy MY-720 tetraglycidyl-4, 4' - methylenedianiline epoxy resin and diaminodiphenyl sulfone (DDS) curing agent. Additional accelerators and/or tackifying resins are added to some of the resin formulations. Since these resin formulations are proprietary, no attempt is made in this paper to reveal the precise composition of any commercial formulation.

The GPC chromatograms of the four epoxy resin formulations analyzed in the uncured state, here designated Resins A through D, are shown in Figures 1 to 4. The composition of the resin formulations are very similar since they are all based upon MY-720 epoxy resin and DDS curing agent. The formulations were doped with MY-720 and DDS to verify the peaks shown in the GPC analyses. For these particular resins the ultraviolet signal and the refractive index signal are equivalent. The ultraviolet signal is reproduced in this paper. Formulation A has a high molecular weight component not present in the other formulations.

The GPC analysis of these uncured epoxy resins gives sharply defined peaks. GPC analysis characteristically gives good resolution in the low molecular weight region. Edwards and Ng[1], for example, found that it is possible with GPC to characterize mixtures of components having molecular weights below a few thousand quite precisely.

This type of GPC analysis can be used for "fingerprint analysis" to detect the major components in the formulation. As will be shown later in this paper, GPC readily detects the presence of higher molecular weight products which form as the resin ages during storage.

However high pressure liquid chromatography (HPLC) operated in the reverse phase mode is superior to GPC in detecting the presence of minor components in the resin. Zucconi and Humphrey[2] and Dark, et al[3] found this to be the case with bisphenol A epoxy resins and it is also true for a tetraglycidyl methylenedianiline epoxy such as MY-720. Figures 5 and 6 present the HPLC analyses of resin formulations B and D. The far greater analytical capability of the HPLC procedure compared to GPC is obvious when Figures 5 and 6 are compared to Figures 2 and 4.

The HPLC analysis of MY-720 resin as received from Ciba-Geigy is shown in Figure 7. There is a rather high level of impurities in the resin. These impurity peaks are present in the epoxy resin formulations as shown in Figures 5 and 6. The effect that these impurities have upon properties of cured epoxy resin is unknown at this time.

The ideal fingerprint analysis would use both GPC and HPLC. GPC analysis shows the major components in the formulation and the molecular size distribution of the components while HPLC shows the presence of minor and trace impurities. Unfortunately reversed phase HPLC has

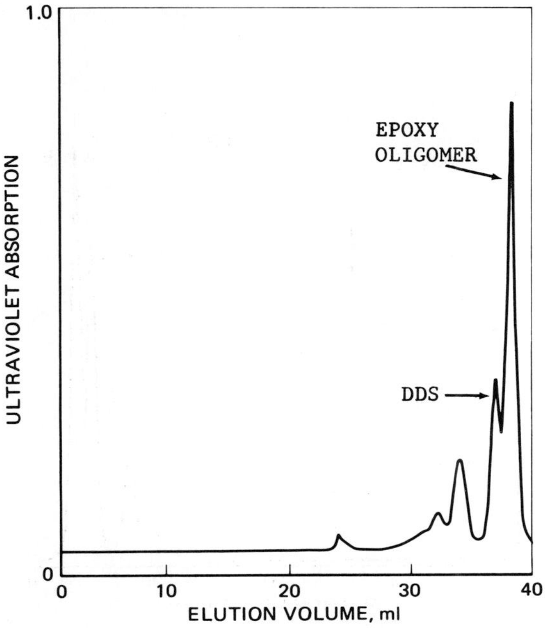

Figure 1. Gel permeation chromatographic analysis of epoxy resin formulation: Formulation A.

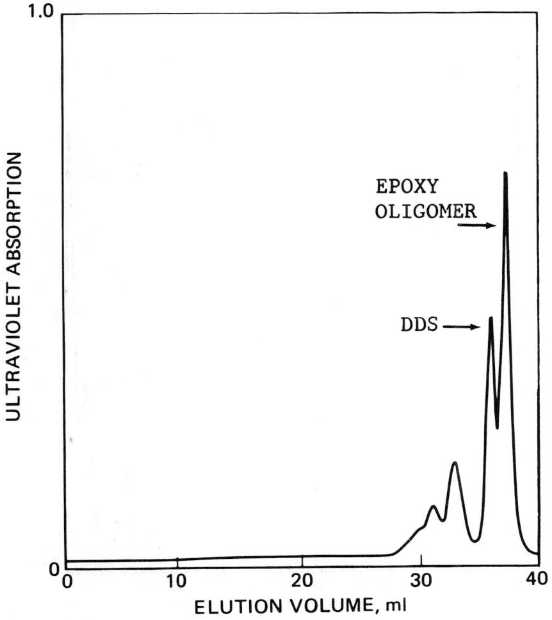

Figure 2. Gel permeation chromatographic analysis of epoxy resin formulation: Formulation B.

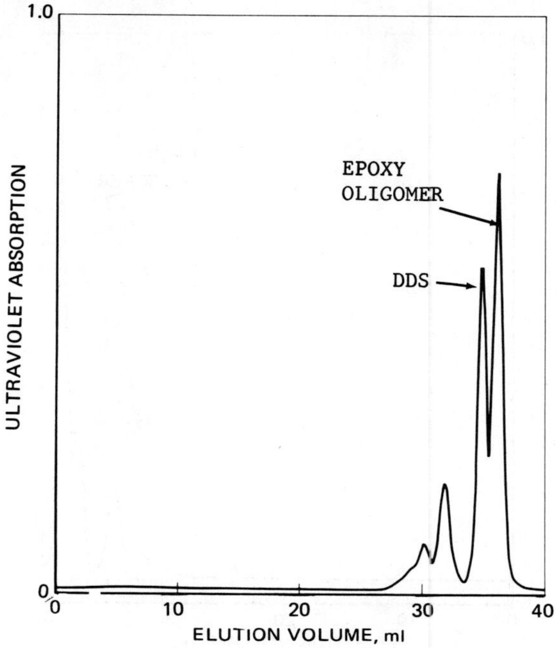

Figure 3. Gel Permeation chromatographic analysis of epoxy resin formulation: Formulation C.

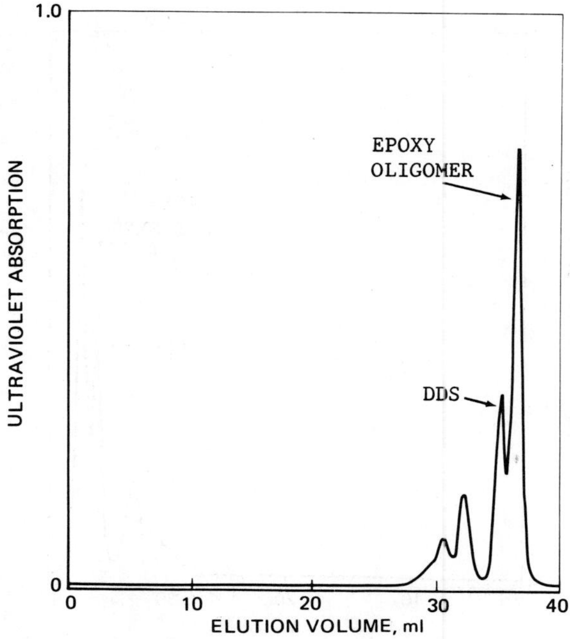

Figure 4. Gel permeation chromatographic analysis of epoxy resin formulation: Formulation D.

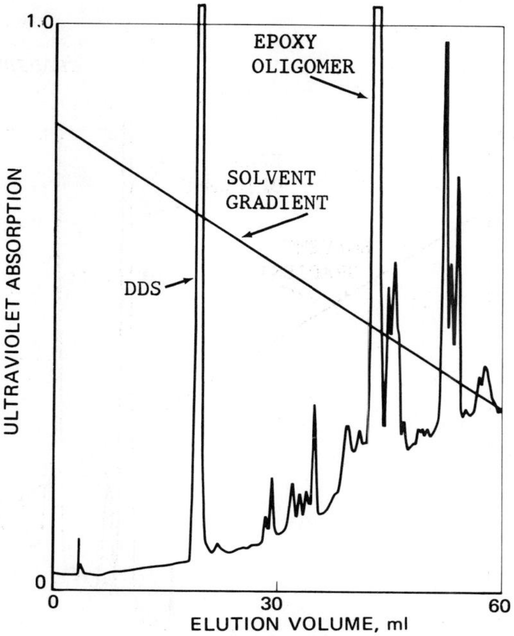

Figure 5. Reversed phase high pressure liquid chromatographic analysis of epoxy resin Formulation B.

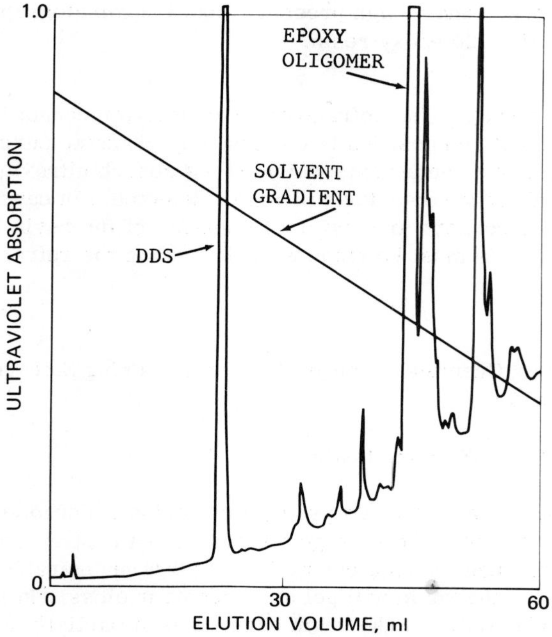

Figure 6. Reversed phase high pressure liquid chromatographic analysis of epoxy resin formulation: Formulation D.

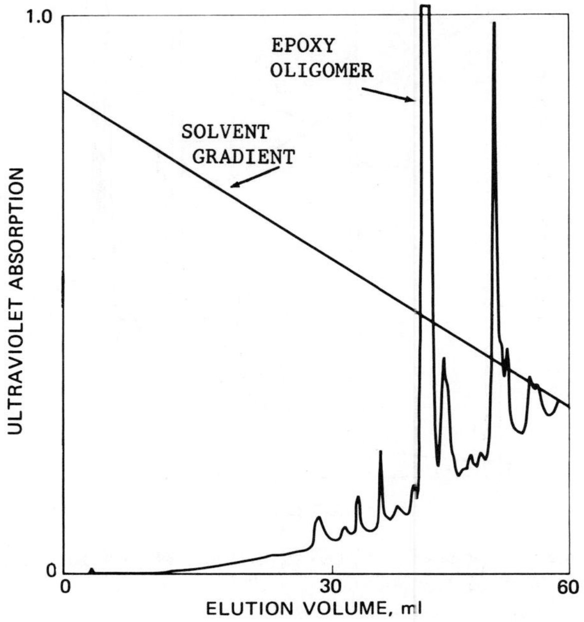

Figure 7. Reversed phase high pressure liquid chromatographic analysis of Ciba-Geigy MY-720 epoxy resin.

one serious drawback. The refractive index detector cannot be used because the solvent composition is continuously changing during the analysis. Therefore components which do not absorb ultraviolet radiation at the wavelength of the detector will not be detected. In cases where non-ultraviolet absorbers are major components of the resin formulation, the primary analysis must be made with GPC using the refractive index detector.

2. Analysis For Chemical Changes Occurring During Storage of Epoxy Formulations

A. MY-720/DDS Epoxy Resin

During storage the catalysed epoxy resins undergo curing at a rate dependent upon the storage temperature and upon the reactivity of the curing agent. This curing is of course undesirable. In the extreme case the resin will gel and become useless. In intermediate cases the chemistry of the resin changes as it partially cures and this introduces an uncertainty in the material that should be avoided.

GPC analyses made on the resin periodically as it ages is an excellent way to track the changes which occur during storage. Eggers and Humphrey[4] used GPC to follow the advancement of brominated bisphenol A epoxy resin catalysed with dicyandiamide. The MY-720 resin containing DDS is a very unreactive system when stored at temperatures from 0—22°C. Figure 8 shows the GPC analysis of Formulation A after 46 days storage at 22°C. Note the gradual reduction in DDS content and growth in higher molecular weight species which is occurring. DDS has a low reactivity compared to other primary aromatic amines such as metaphenylenediamine. Figure 9 shows the gel permeation chromatographic analysis of Formulation B after 1 year at 22°C. Here the growth of higher molecular weight species and disappearance of DDS is extreme.

The reactivity of the MY-720/DDS system is very low at 0°C. Figures 10 and 11 show the GPC analyses of 2 lots of Formulation C after 3 years storage at 0°C. Higher molecular weight species have not formed nor has there been an obvious reduction in DDS content. It was totally unexpected that Formulation C would be this unreactive. To verify that no changes had occurred in the resin, graphite fiber composites were prepared from the two lots of the formulation and tested for interlaminar shear strength and flexural strength. The composite strengths were identical to the strength values obtained three years earlier when the resin was first formulated.

B. Dow 7818/Epon Z Epoxy Resin

Not all epoxy resins of importance to the aerospace industry are based upon MY-720/DDS. One epoxy formulation that is being considered for the field repair of damaged composite structures is a blend of Dow 7818 bisphenol F epoxy resin with Shell Epon Z curing agent. This formulation is attractive because it makes use of an easily used liquid curing agent and a low viscosity bisphenol F epoxy resin. This formulation is far more reactive than MY-720/DDS. In a 120 gram mass with 20 parts Epon Z per hundred parts resin the formulation has a pot life of about 24 hours at 22°C. GPC detects the changes which occur in the resin as it gels at this temperature and also the changes which occur after gellation as long as the resin is soluble.

Figure 12 shows the GPC analysis of 7818/Epon Z as freshly mixed. Epon Z is a mixture of primary aromatic amines which are much more reactive than DDS. After 24 hours at 22°C the 120 gram mass of resin has gelled. It is still soluble however and GPC analysis (Figure 13) shows most of the original components have reacted leading to the formation of higher molecular weight species. After four

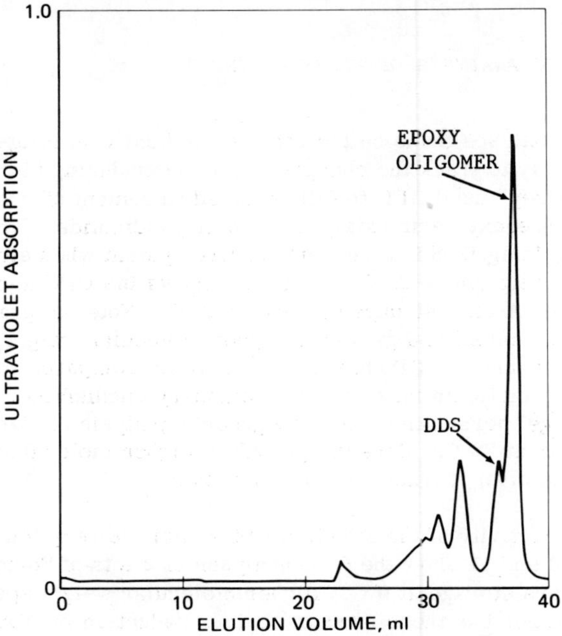

Figure 8. Gel permeation chromatographic analysis of Formulation A after 46 days storage at 22°C.

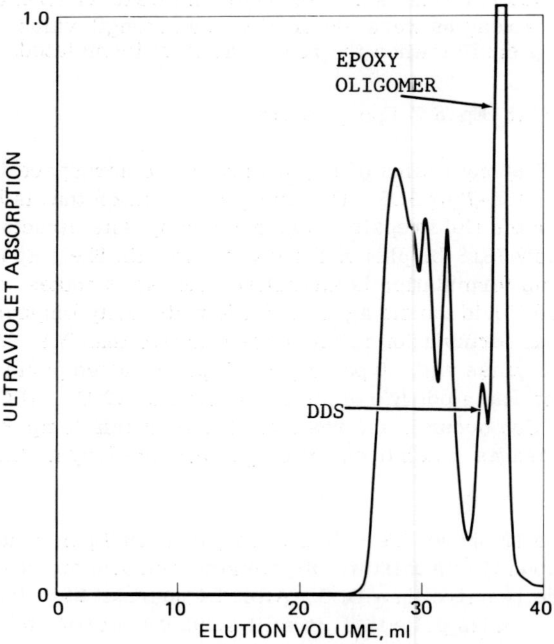

Figure 9. Gel permeation chromatographic analysis of Formulation B after one year storage at 22°C.

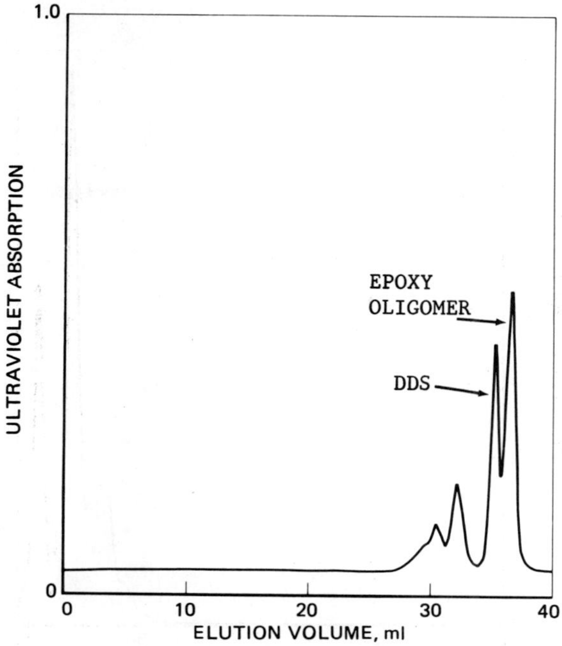

Figure 10. Gel permeation chromatographic analysis of Formulation C Lot 1 after 3 years storage at 0°C.

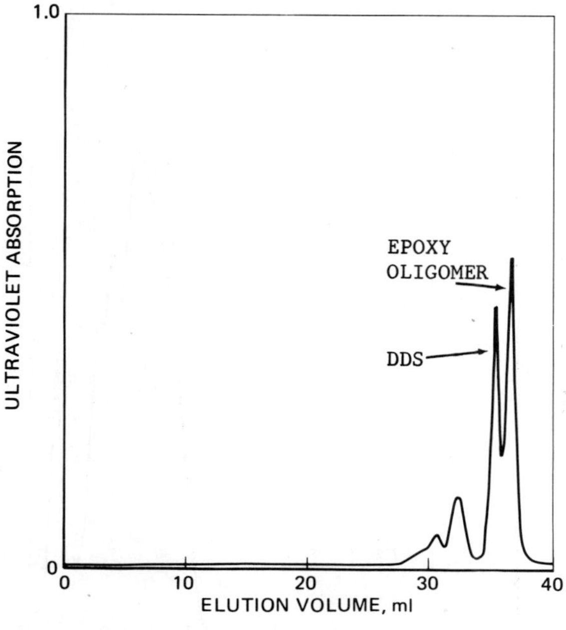

Figure 11. Gel permeation chromatographic analysis of Formulation C Lot 2 after 3 years storage at 0°C.

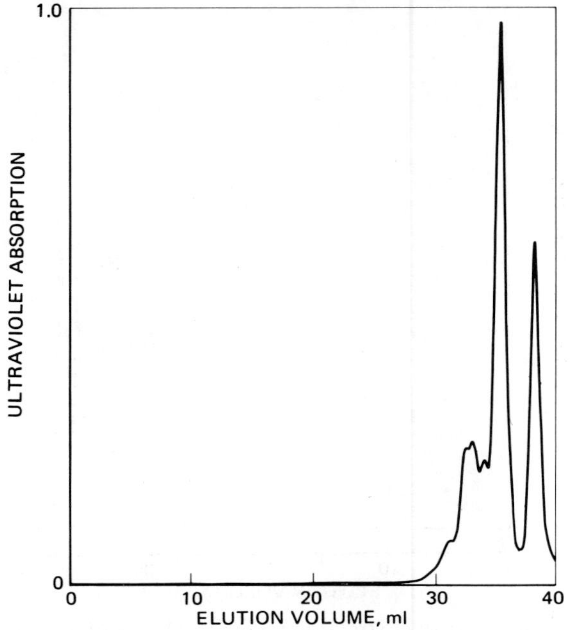

Figure 12. Gel permeation chromatographic analysis of freshly mixed 7818/Epon Z epoxy resin formulation.

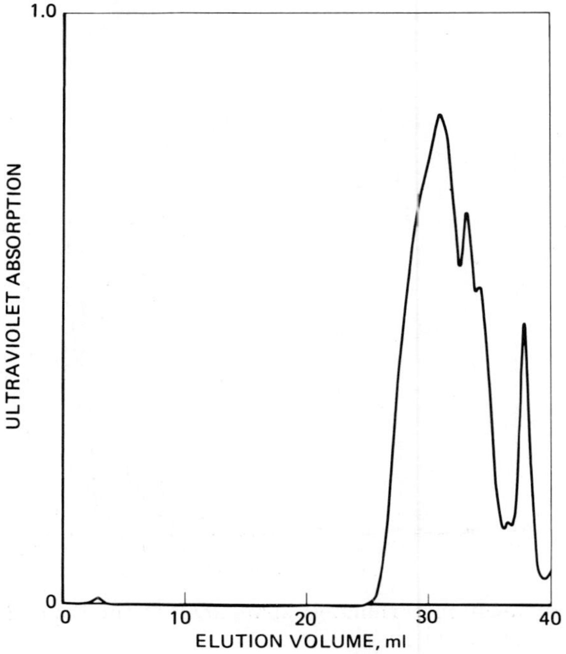

Figure 13. Gel permeation chromatographic analysis of 7818/Epon Z epoxy resin formulation 24 hours after mixing.

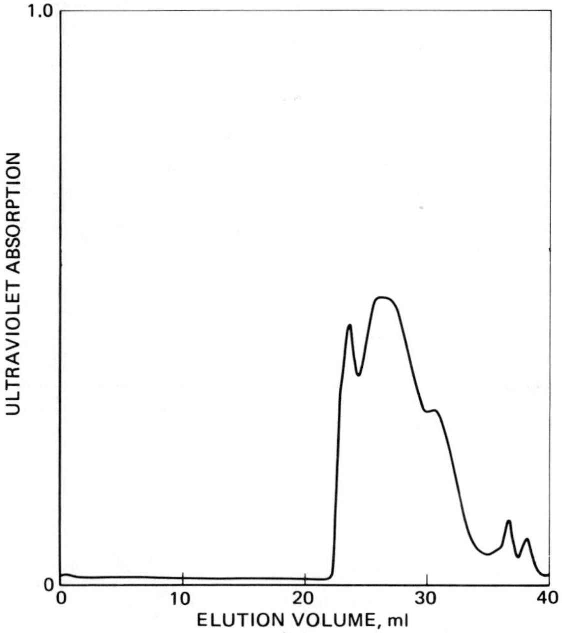

Figure 14. Gel permeation chromatographic analysis of 7818/Epon Z epoxy resin formulation 4 days after mixing.

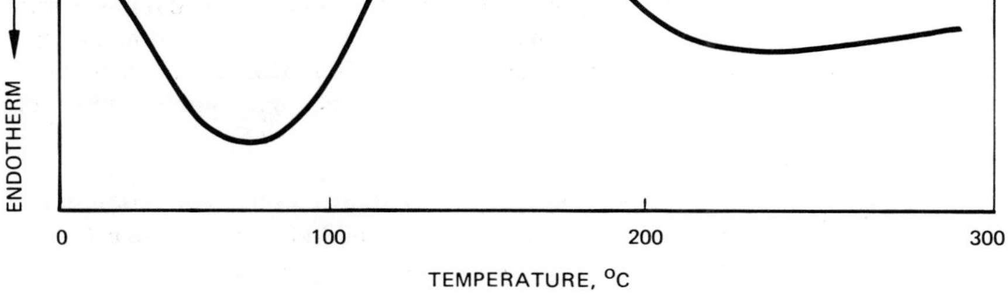

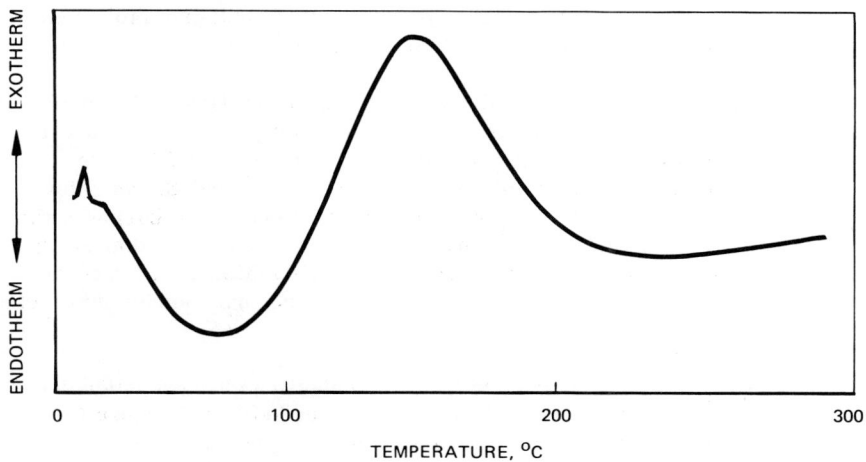

Figure 15. Differential thermal analysis of 7818/Epon Z epoxy resin formulation after 24 hours at 22°C.

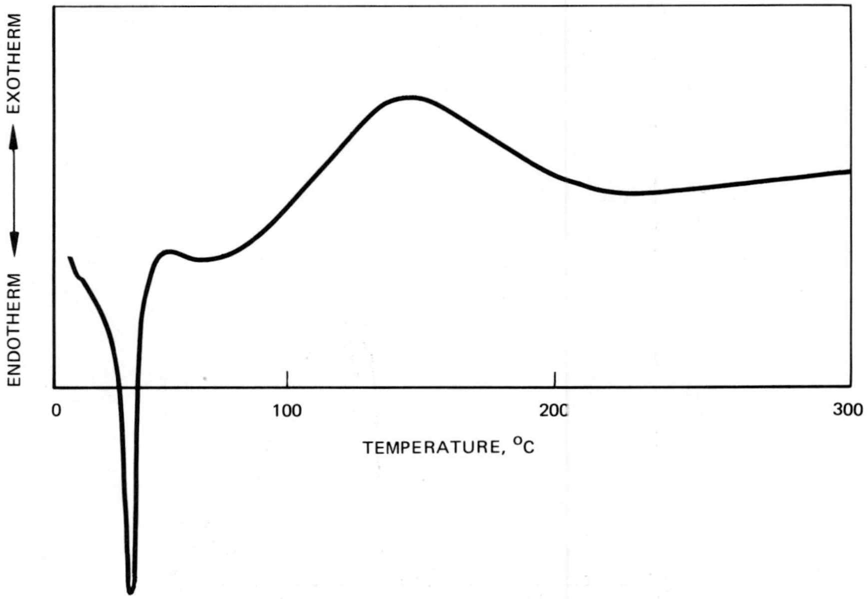

Figure 16. Differential thermal analysis of 7818/Epon Z epoxy resin formulation after 8 days at 22°C.

days at 22 C GPC (Figure 14) shows that only traces of the starting materials remain. Much of the resin is now of such high molecular weight that it is totally excluded from these columns.

Differential thermal analysis shows that considerable amounts of unreacted epoxide remains in the resin after gellation. Figure 15 shows the differential thermal analysis of 7818/Epon Z 24 hours (resin held at 22 C) after mixing. A broad diffuse endotherm below 100 C is followed by an exotherm above 100 C due to the curing of the epoxy. Differential thermal analysis eight days (resin held at 22 C) after mixing (Figure 16) shows the now brittle material has a sharp melting point near 50 C. The exotherm still occurs showing unreacted epoxide is still present.

GPC analyses of this type are ideal for detecting changes which occur in the resin during storage. HPLC is not needed here because GPC readily detects the growth of higher molecular weight species which is the major change that occurs during storage. Without GPC analysis a composite would have to be molded from the material in question and tested. GPC analysis can be made far more economically.

CONCLUSIONS

1. Analyses of N, N, N', N' - tetraglycidyl -4, 4' - methylenedianiline epoxy resin formulations catalysed with diaminodiphenyl sulfone curing agent and bisphenol F epoxy resin catalysed with primary aromatic amines can be readily analyzed using GPC and HPLC. The HPLC analysis is much more sensitive than GPC; however, HPLC is limited in that only the ultraviolet detector can be used. GPC either used alone or used with HPLC makes an excellent analytical method for determining the composition of epoxy resin formulations which are purchased commercially.

2. GPC readily detects the changes which occur in the chemical composition of catalysed epoxy resins during storage at room temperature and at reduced temperature. The disappearance of aromatic primary amine curing agents and the development of higher molecular weight components can be followed. An analysis of this type is of considerable practical importance because aircraft structures which are critical to flight safety are being manufactured from these epoxy resin formulations.

REFERENCES

1. G. D. Edwards and Q. Y. Ng, J. Polymer Sci, C, 21, 105 (1968).

2. T. D. Zucconi and J. S. Humphrey, Jr., Soc. Plast. Eng. Tech. Pap., 20, 595 (1974).

3. W. A. Dark, E. C. Conrad, and L. W. Crossman, Jr., J. Chromatogr. 91, 247 (1974).

4. E. A. Eggers and J. S. Humphrey, Jr., J. Chromatogr., 55, 33 (1971).

EFFECT OF DIFFERENT CATALYSTS ON AN IDENTICAL THERMOSETTING EPOXY-ANHYDRIDE RESIN SYSTEM EVALUATED THROUGH THE B-STAGED CURING CONDITIONS BY GEL PERMEATION CHROMATOGRAPHY

R. D. Nuss

Brunswick Corporation
Lincoln, Nebraska

ABSTRACT

This preliminary investigation involves the gel permeation chromatography's capability of evaluating thermosetting epoxy-anhydride resin systems through the B-stage curing cycle. The results demonstrate the microscopic polymerization differences, which can occur due to different catalysis. The important role of the catalyst was illustrated by the apparent molecular size distribution differences between the identical resin systems at various time intervals throughout the thermosetting epoxy-anhydride cure process.

I. INTRODUCTION

Through the last few years, there have been many applications for gel permeation chromatography (GPC) in the polymer industry including the analyzation of thermosetting epoxy resin systems. Most thermosetting epoxy resin work has been involved with the characterization of epoxy resins [1-3]. One exception was the report by Humphrey and Eggers (1971), who demonstrated the application of gel permeation chromatography in the evaluation of B-stage prepreg epoxy-glass printed circuit laminates [4]. This paper deals also with the evaluation of B-stage epoxy resins by gel permeation chromatography; however, the evaluation involves analyzing two epoxy-anhydride catalysts in an identical thermosetting epoxy-anhydride resin system as they advance through a B-stage curing cycle. The results demonstrate the important controlling role and effect a catalyst has on a thermosetting epoxy-anhydride resin system.

The basic epoxy-anhydride reaction is: "The anhydrides react by combining with both the epoxy groups and the hydroxyl groups present in most epoxy resins. The reaction is initiated by the formation of the half ester through the anhydride joining the hydroxyl group. The remaining acid group then reacts with an epoxy group. In addition, self-polymerization of the epoxy group through the epoxy-hydroxyl reaction may occur. So, both ester and ether linkages can be created········. The favored reaction is dependent upon the curing temperature, accelerator used, etc. [5]" [6].

II. EXPERIMENTAL CONDITIONS

The two evaluated catalysts were catalyst A (a tertiary amine catalyst) and catalyst B (a proprietary catalyst). This investigation involved determining what effect these catalysts have on an identical thermosetting epoxy-anhydride resin system. The two resin systems evaluated have the following resin formulas as shown:

RESIN SYSTEM #1		RESIN SYSTEM #2	
Components	Parts By Resin Wt.	Components	Parts By Resin Wt.
Bis A-epoxy	100	Bis A-epoxy	100
Anhydride	80	Anhydride	80
Catalyst A	1	Catalyst B	1

As noticed, the two resin systems differ only in the type of catalyst used (the variable factor in this investigation).

The following test conditions were established. First, a correct relationship epoxy-anhydride mixture was divided into two equal 520 gram portions: one portion was mixed with catalyst A and the other portion was mixed with catalyst B. This mixing procedure ensured two identical relationship epoxy-anhydride resin systems. Second, these two 520 gram portions were transferred to a Blue M electric oven (Model CW-1604 with 150°F - 800°F temperature range) for the desired cure cycle. The cure cycle involved: (1) 8-1/2 hours at 150°F; (2) 2 hours at 200°F; and (3) 2 hours at 300°F. Third, before and during the cure cycle, at various time intervals, 0.200 gram resin samples were taken, which were weighed into a 250 ml glass stopper flask. After the resin samples were weighed out, 50 ml of THF (tetrahydrofuran) solvent was added immediately to interrupt further polymerization activity. These THF resin samples, representing the B-stage advancement of both thermosetting epoxy-anhydride resin systems, were later evaluated on a Waters Associates Model ALC/GPC 201. This instrument consists: (1) a single 2,000 psi pump; (2) universal injector with a 0.5 ml sample size injection; and (3) a refractive index detector. The GPC operating parameters were as follows: (1) solvent was THF; (2) temperature was 75°F; (3) flow rate was 1.3 ml/min; (4) sensitivity settings were 2X to 16X; and (5) columns were 3-100Å and 2-500Å μ-styragel columns. A Texas Instrument recorder was utilized to record the refractive index output, and its chart speed was 0.75 in/min.

III. RESULTS AND DISCUSSION

Figure 1 illustrates the characteristic output spectrum from the Model ALC/GPC 201 instrument, showing the molecular size distribution and position of Bis A-epoxy, anhydride, catalyst A, and catalyst B in relationship to the injection point. The number in the parenthesis indicates the amplification intensity setting, which was regulated to obtain peak height optimization for that specific peak; therefore, the peak height concentration order (determined by multiplying the peak height by the amplification intensity) is: Bis A-epoxy peak ($n = 0$ oligomer) > anhydride peak > catalyst A peak > catalyst B peak > polymerization peak ($n = 1$

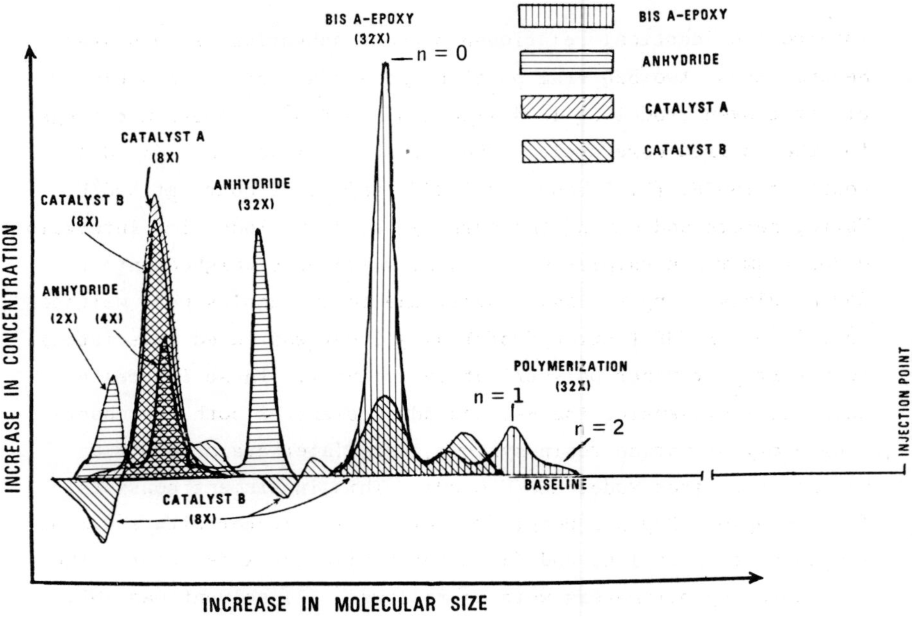

Figure 1. The characteristic output spectrum from Model ALC/GPC 201 instrument showing the molecular size distribution and position of resin system components.

oligomer of Bis A-epoxy)[2]. The important observation is the molecular size distribution and position of the two catalysts in relationship to each other, and to Bis A-epoxy and anhydride peaks. As observed, catalyst B has a broad molecular size distribution (at least seven component species consisting of two negative peaks) in which the first two peaks from the left of the injection point correspond to the n = 0 oligomer and impurity peak of Bis A-epoxy [2]. The catalyst B also has molecular size material similar to both the catalyst A and the middle anhydride peak material. The combination of the Bis A-epoxy, anhydride and catalyst spectra for each resin system would produce two similar thermosetting epoxy-anhydride resin system spectra. Figure 2 compares the two freshly mixed thermosetting epoxy-anhydride resin system spectra. (Note

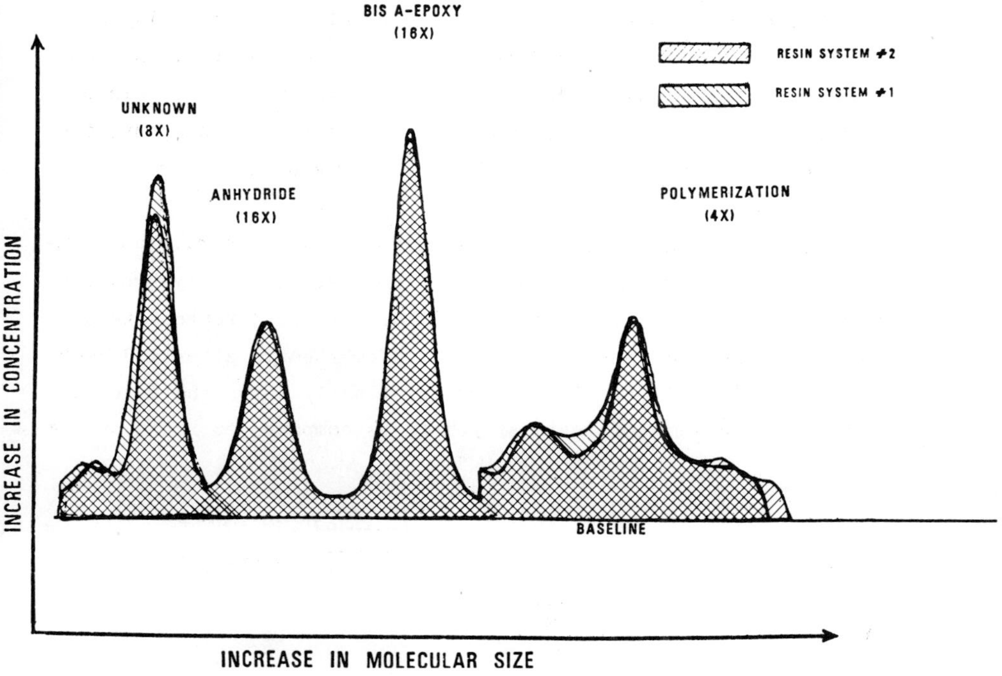

Figure 2. The freshly mixed thermosetting epoxy-anhydride resin system #1 and #2 spectra.

that a smaller amount of catalyst was incorporated into the resin formulation in Figure 2 as compared to Figure 1, which represents just the catalyst spectrum itself.) As expected, the resin system spectra are similar. However, subtle differences do exist between the resin systems in three categories, relating to the catalyst concentration and effect on the thermosetting epoxy-anhydride systems. First, the concentration difference of Bis A-epoxy peak is due to the small amount of catalyst B present in that molecular size position. Second, the concentration difference of unknown peak (assumed to be a combination of catalyst and a fractional portion from the anhydride) is related to the concentration difference between catalyst A and catalyst B in that molecular size position. Third, the comparison of the Bis A-epoxy oligomer polymerization peaks illustrates resin system #1 has a less distinct oligomer distribution than resin system #2, whereas resin system #2 has a broader oligomer material than resin system #1. Since both resin systems have the same amount of anhydride present, the identical anhydride concentration level between the two resin systems is expected.

The succeeding figures and discussion will illustrate and compare the B-stage advancement of both thermosetting epoxy-anhydride resin systems at various time intervals. Figure 3 represents both thermosetting epoxy-anhydride resin system spectra after 1.0 hour of B-staging at 150°F. This figure distinctly shows the rate of polymer advancement of resin system #2 as compared to resin system

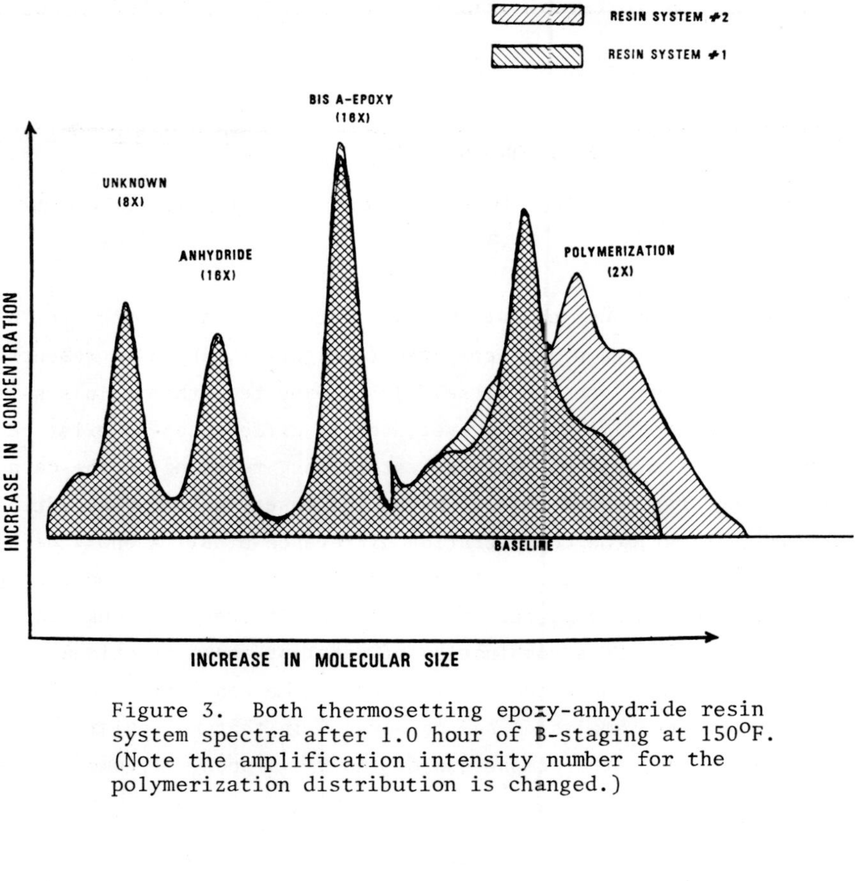

Figure 3. Both thermosetting epoxy-anhydride resin system spectra after 1.0 hour of B-staging at 150°F. (Note the amplification intensity number for the polymerization distribution is changed.)

#1 (which advanced little after one hour). In fact, resin system #1 has a similar basic polymerization peak distribution as the freshly mixed resin system #1 except for a slight concentration level increase (note the amplification intensity number is different in this polymerization distribution); whereas, resin system #2 has changed remarkably in the polymerization distribution, which has increased and produced at least two larger molecular size (oligomers) components, as compared to the freshly mixed resin system #2. Apparently, this increased coupling of oligomers (polymerization peaks) is the result of the catalyst and probably due in part to the reactive hydrogen present in the polymerization components (n = 1 and n = 2 oligomers) but absent in the two monomers (anhydride and n = 0 oligomer). Thus, the polymerization distribution increase of resin system #2 is directly related to the decreased Bis A-epoxy concentration level, because Bis A-epoxy (n = 0 oligomer) material is one of the major building block elements for the polymerization material. This suggests that catalyst B has a different effect on a thermosetting epoxy-anhydride resin system than the catalyst A, especially in its epoxy-catalyst reaction site and its rate of reactivity. Figure 3 also shows the unknown and anhydride peaks at the same concentration levels, indicating the same incorporation rate of these elements into the polymerization material.

Figure 4 represents the next comparison stage of the two resin systems after 3.8 hours of B-staging. As noticed, there still remains a dramatic difference between these two resin systems, especially in the polymerization distributions. The polymerization distributions of both resin systems have shifted to larger molecular size distributions, and the original polymerization distributions either have disappeared or are decreasing in concentration. The polymerization distribution of resin system #2, which has developed an extremely larger molecular size component and also indicates the exclusion volume position of these specific columns, is still advancing faster than the polymerization distribution of resin system #1, illustrating the indirect catalyst effect on the

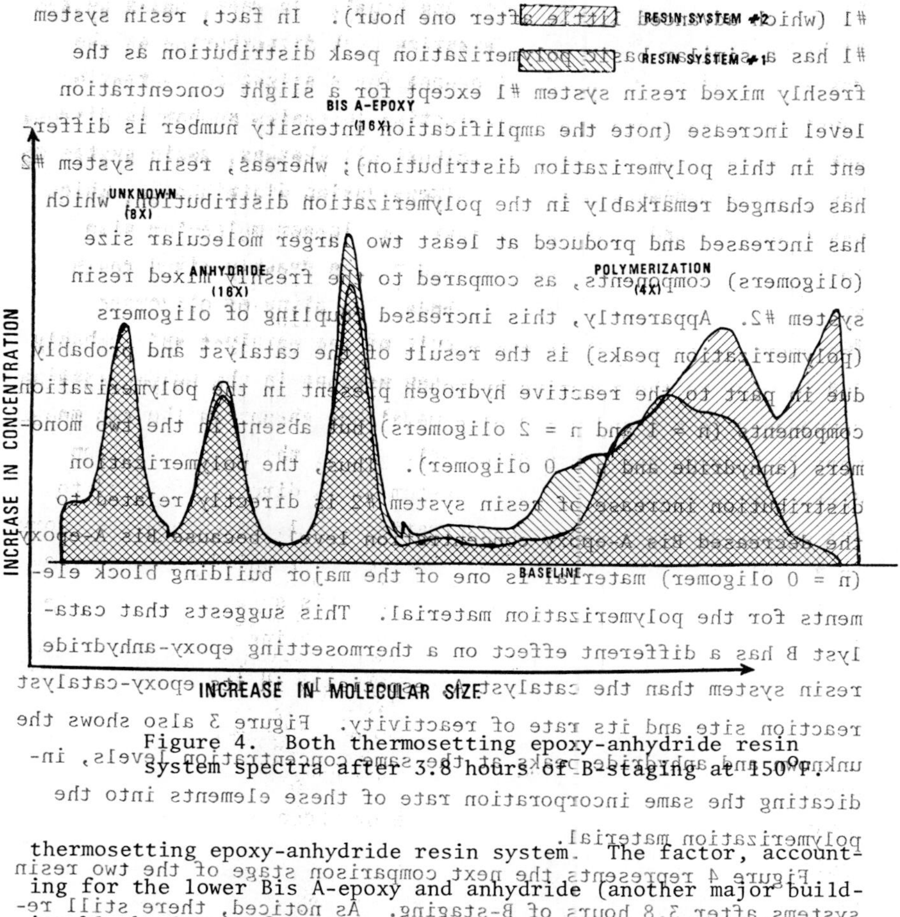

Figure 4. Both thermosetting epoxy-anhydride resin system spectra after 3.8 hours of B-staging at 150°F.

thermosetting epoxy-anhydride resin system. The factor, accounting for the lower Bis A-epoxy and anhydride (another major building block element for polymerization material) concentration levels of resin system #2 as compared to the resin system #1, is the polymerization rate of the resin system. The unknown concentration level for both resin systems is equal but at a lower concentration level than the 1.0 hour spectra. This would indicate that either the catalysts or the middle anhydride peak material are/is being incorporated at an equivalent rate into the polymerization reaction of both resin systems.

Figure 5 represents the two thermosetting epoxy-anhydride resin system spectra after 6.4 hours of B-staging, which were taken just prior to the resin systems becoming solidified. At this

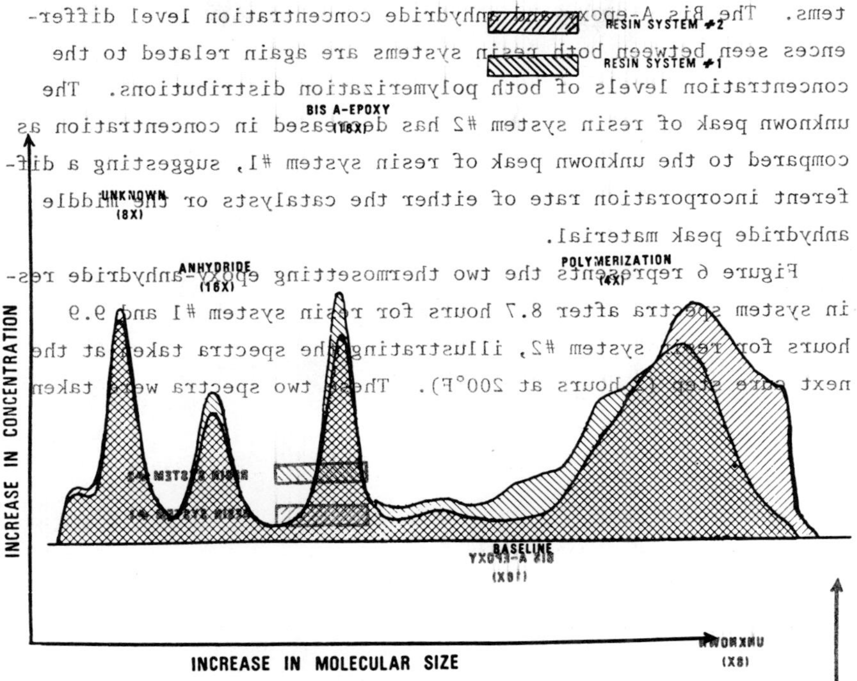

Figure 5. Both thermosetting epoxy-anhydride resin system spectra after 6.4 hours of B-staging at 150°F.

time interval, the polymerization distributions of both resin systems are becoming similar in concentration and distribution, indicating the polymerization distribution advancement of resin system #1 is proceeding at an equivalent rate to the polymerization distribution of resin system #2. This factor is supported by comparing Figures 4 and 5, illustrating that the polymerization distributions of both resin systems have shifted little to larger molecular size material. However, the concentration levels of both polymerization distributions have increased except for the largest molecular size material of resin system #2. The disappearance of the largest molecular size material is caused by the material advancing into an insoluble THF state and thus being filtered out of solution. This represents the first visible sign of detectable insoluble sample material in THF from either resin systems. The Bis A-epoxy and anhydride concentration level differences seen between both resin systems are again related to the concentration levels of both polymerization distributions. The unknown peak of resin system #2 has decreased in concentration as compared to the unknown peak of resin system #1, suggesting a different incorporation rate of either the catalysts or the middle anhydride peak material.

Figure 6 represents the two thermosetting epoxy-anhydride resin system spectra after 8.7 hours for resin system #1 and 9.9 hours for resin system #2, illustrating the spectra taken at the next cure step (2 hours at 200°F). These two spectra were taken

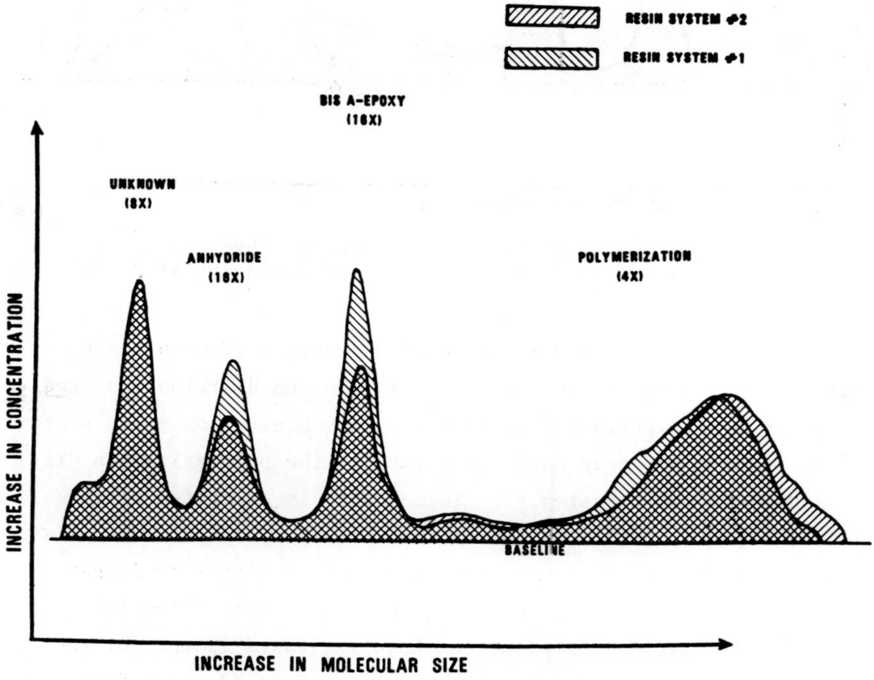

Figure 6. Both thermosetting epoxy-anhydride resin system spectra after 8.7 hours for resin system #1 & 9.9 hours for resin system #2. These spectra illustrate the resin components left after gelation.

after the gelation stage, representing the remaining resin system material which is soluble in THF. As noticed, the polymerization distributions of both resin systems have decreased in concentration and have a reduced distribution size. This disappearance of polymerization material indicates the incorporation of this material into building a solid resin matrix which, at this stage, has completed a certain percentage of the total crosslinked structure. The Bis A-epoxy, anhydride, unknown, and polymerization concentration levels of both resin systems in Figure 6 indicate the remaining resin system components to be incorporated into the crosslinked resin matrix. Thus, these spectra suggest the possibility of following a thermosetting epoxy resin system through the full cure cycle by evaluating the unreacted resin system components.

Figures 7 and 8 represent the rate of disappearance of Bis A-epoxy and anhydride, respectively, and illustrate the approxi-

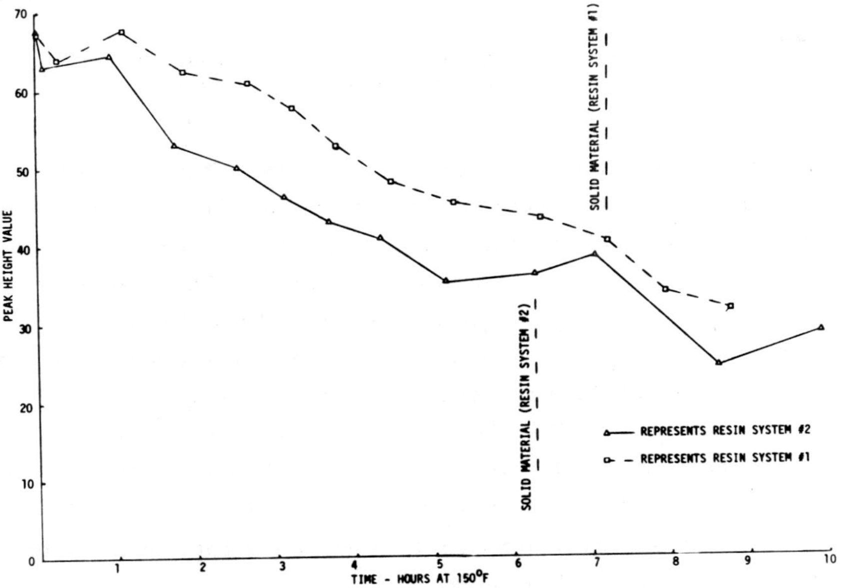

Figure 7. The rate of Bis A-epoxy disappearance through the cure cycle for both resin systems.

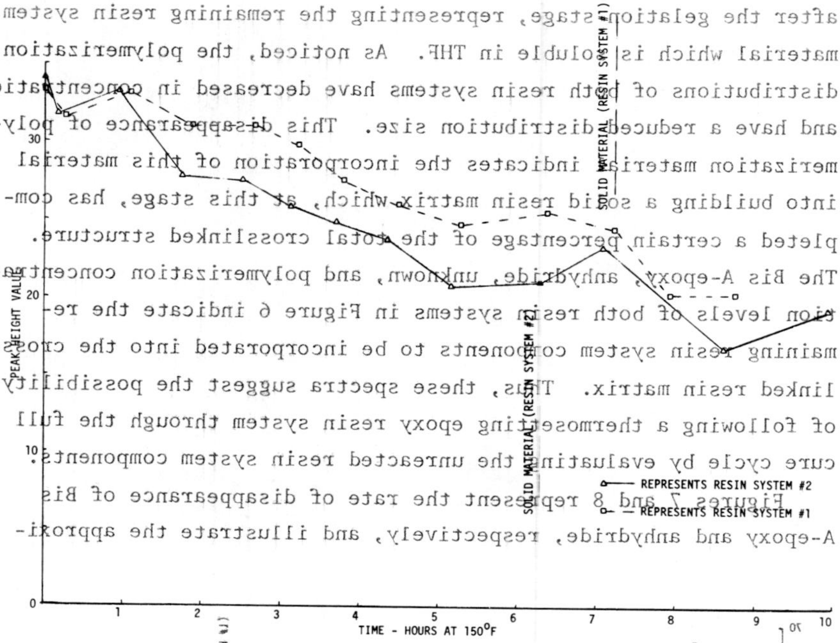

Figure 8. The rate of anhydride disappearance through the cure cycle for both resin systems.

mate gelation position of both resin systems. These figures reinforce the indirect effect the catalyst has on the resin system reactivity by controlling the reaction rate of the two major resin system components (Bis A-epoxy and anhydride). The incorporation of catalyst B into a thermosetting epoxy resin system results in a faster reacting B-staged resin system than catalyst A through the first six hours; whereas, the same incorporation rate of Bis A-epoxy and anhydride for both resin systems occurred after the first 7 hours of the B-staged cure cycle. This indicates that either the catalysts no longer affect the reaction rate or the catalysts affect the reaction rate equally for both resin systems after seven hours of B-staging.

IV. SUMMARY

This preliminary information indicates the catalyst is the controlling factor in the development of thermosetting epoxy-anhydride resin system. The investigation showed what a small amount of catalyst can do to affect the polymerization distribution and the incorporation rate of Bis A-epoxy and anhydride into a thermosetting epoxy-anhydride resin system. The results indicate resin system #2 (catalyst B) advanced faster and developed larger polymerization (molecular size) material at the first six hours of B-staging than the resin system #1 (catalyst A); whereas, the latter stage of the B-staging process indicate the two resin systems becoming similar in polymer material.

Another important aspect of this effect is the illustration of the capability of gel permeation chromatography in evaluating and studying the B-stage cycle of thermosetting epoxy-anhydride resin systems including a suggestion of the possibility of evaluating the complete cure cycle.

ACKNOWLEDGEMENTS

Acknowledgement is made to Dr. Delmar Timm, Professor of Chemical Engineering at University of Nebraska for providing the gel permeation chromatography equipment and consultation.

REFERENCES

1. G. D. Edwards and Q. Y. Ng, J. Polym. Sci., Part C (1968).
2. F. N. Larsen, 6th Intern. Seminar of Gel Permeation Chromatography, Miami Beach, Fla., October (1968).
3. G. H. Miles, National ACS Meeting, Chicago, Ill., (1965).
4. E. A. Eggers and J. S. Humphrey, Jr., J. Chromatog., 55 (1971).
5. W. Fisch and W. Hufmann, Plastics Technol., August (1961).
6. George Lubin, Handbook Of Fiberglass And Advanced Plastic Composites, (1969).

*THE GEL PERMEATION CHROMATOGRAPHY OF OLIGOMERS**

M. R. Ambler and R. D. Mate

Chemical Materials Development
The Goodyear Tire and Rubber Company
Akron, Ohio

INTRODUCTION

In the gel permeation chromatography (GPC) of high molecular weight polymers, $M[\eta]$ has been found by several authors to be a "universal calibration" parameter (Refs. 1, 2). For Gaussian coils, the mechanism governing the GPC separation involves the hydrodyamic volume, which is proportional to $M[\eta]$. For the GPC of low molecular weight compounds, molecular volume appears to be the governing factor for the separation of a number of alkanes and aromatic compounds (Ref. 3). For the intermediate molecular weight range, one article recently published by Belenkii and coworkers reported that several types of polyester oligomers followed a common $M[\eta]$ calibration curve (Ref. 4). This communication relates our findings regarding the GPC of several types of oligomeric polymers.

Experimental

A Waters Associates Model 200 GPC equipped with a differential refractometer detector was used with toluene at 30° and a flow rate of 1 cc/min. The Styragel column set consisting of five-four foot long columns of porosity ratings 10^3, 350-700, 10^2, 10^2, 10^2A had a plate count of 1160 plates/foot. The plot of log M versus elution volume for a series of linear polystyrene standards was linear over the molecular weight range from 106 to around 10,000 g/mole. Column resolution was lost for higher molecular weights and the calibration curve showed upward curvature. To generate the calibration curves, the peak elution volume (PEV) was assigned the values of $M[\eta]$ and M. Intrinsic viscosities were determined with a capillary viscometer having a shear rate of 500 sec^{-1} for toluene. Kinetic energy corrections were applied to

---

the data. Solutions with relative viscosities (η rel) as low as 1.1 were needed to obtain linear plots of (ηsp/c) versus c and ln(η rel/c) versus c.

Oligomers used in this study included polystyrene (PSty), polybutadiene (PBd), polyisoprene, (PIso), hydrogenated polyisoprene (h-PIso), polyethylene (PE) and poly (propylene oxide) glycol (PPOG). The PSty and PPOG samples were those commercially available from Waters Associates. The PBd and PIso samples were anionically polymerized in the laboratories of The Goodyear Tire and Rubber Company. The remaining oligomeric and low molecular weight PSty, h-PIso and PE samples were obtained as reagent grade chemicals from several chemical supply houses.

Results

All molecular weight values are listed in Table I. The molecular weights assigned to the commercial PSty, PPOG, PE and h-PIso samples were those furnished by the suppliers. Characterization of the remaining samples included vapor pressure osmometry, membrane osmometry and the average molecular weight calculated from the kinetics of their anionic polymerization. The molecular weight distributions (MWD) for these samples are described in Table II by the half-widths of their GPC chromatograms (see the insert in Figure 4). These data indicated that (except for the PPOG samples) the various samples employed were approximately as narrow as the commercial PSty standards, i.e., ($\bar{M}w/\bar{M}n$) ≤ 1.1. Skewing of the distributions also appeared to be minimal. Therefore, assignment of the peak elution volume (PEV) to the molecular weight and intrinsic viscosity of the sample appeared justified. Figure 1 shows GPC traces of some of the samples as they were eluted from a GPC column set similar to the one used here. Not all of the samples are shown here, but these serve to illustrate the monodispersity of these samples.

In most cases, intrinsic viscosities were obtained experimentally. For others, values of [η] were interpolated from the appropriate Mark-Houwink relationships for oligomers compiled by Bianchi and Peterlin (Ref. 5), and the data of Beattie and Booth for polyisoprene (Ref. 6). Figure 2 shows the plots of log [η] versus log $\bar{M}w$ for the various polymer types studied here. Obvious differences in the solution properties are evident. In particular, for most samples (e.g., polystyrene), distinct differences in slope exist at different molecular weight levels. A brief explanation of the cause of this behavior follows, since the basis for these slope changes are related to the problems that are inherent in the GPC analyses of oligomers. The experimental line for polystyrene best exemplifies the theoretical relationship predicted for oligomers (Ref. 5). For high molecular weight Gaussian chains, a slope of about 0.7 in good solvents is expected, but as the molecular weight decreases and the chain length becomes too short for it to obey Gaussian statistics in its

TABLE I

Characterization of Samples

Polymer Type(a)	Sample	$\overline{M}w$(LS)(b)	$\overline{M}n$(VPO)(b)	$\overline{M}n$(MO)(b)	$\overline{M}k$
PSty	25168	20,800	20,200	-	-
	25171	10,000	9,600	10,000	-
	25169	4,000	3,100	4,000	-
	26971	2,100	1,950	-	-
	Ethylbenzene	106	-	-	-
PBd	1483	-	-	-	2940
	1482	-	2,710	-	2780
	1484	-	-	-	2640
	197-H		960	-	904
	197-L	-	440	-	456
PIso	118	-	-	-	29,000
	117		-	-	22,000
	891	-	940	-	895
h-PIso	Squalane	423	-	-	-
PE	Eicosane	282	-	-	-
	Dodecane	170	-	-	-
PPOG	41983	-	3900	-	-
	41985	-	2020	-	-
	41994	-	1220	-	
	41993	-	790	-	-

(a) PSty-polystyrene
PBd-polybutadiene
PIso-polyisoprene
h-PIso-hydrogenated polyisoprene
PE-polyethylene
PPOG-poly(propylene oxide) glycol

(b) LS - light scattering.
VPO - vapor pressure osmometry.
MO - membrane osmometry

Polymer Type (a)	Sample	[η], dl/g Calcd	XPTL	M[η]
PSty	25168	0.147	-	3058.
	25171	0.088	-	880.
	25169	0.055	-	220.
	26971	0.040	-	84.0
	Ethylbenzene	-	0.00145	0.154
PBd	1483	0.117	-	344.
	1482	0.113	-	314.
	1484	0.110	-	290.
	197-H	-	0.052	47.0
	197-L	-	0.030	13.7
PIso	118	0.360	-	10,440.
	117	0.290	-	6,380.
	891	-	0.037	33.5
h-PIso	Squalane	-	0.022	9.09
PE	Eicosane	0.024	-	6.68
	Dodecane	0.011	-	1.82
PPOG	41983	0.052	-	104.
	41985	-	0.040	48.8
	41994	0.032	-	25.4
	41993			

TABLE II

GPC Analysis of Samples

Polymer Type	Sample	PEV, ct	$W_{1/2}$, ct (a)	W_h, ct (a)	W_l, ct (a)	W_l/W_h (a)
PSty	25168	50.36	-	-	-	-
	25171	51.67	-	-	-	-
	25169	54.86	3.48	1.52	1.96	1.29
	26971	57.38	4.12	1.84	2.28	1.24
	Ethylbenzene	87.39	1.72	0.78	0.94	1.20
PBd	1483	53.47	2.85	1.29	1.56	1.21
	1482	53.73	2.69	1.30	1.39	1.07
	1484	53.65	2.75	1.31	1.44	1.10
	197-H	58.38	5.08	2.46	2.62	1.06
	197-H	62.48	3.97	1.82	2.15	1.18
PIso	118	50.15	-	-	-	-
	117	50.34	-	-	-	-
	891	59.00	4.57	2.12	2.44	1.14
h-PIso	Squalane	64.84	1.52	0.68	0.84	1.24
PE	Eicosane	68.28	2.34	1.24	1.10	0.89
	Dodecane	74.28	0.84	0.47	0.37	0.80
PPOG	41983	64.00	9.85	2.42	7.43	3.07
	41985	70.00	10.07	2.60	7.47	2.87
	41994	74.38	11.19	3.13	8.06	2.58
	41993	81.28	13.68	4.21	9.47	2.25

(a) See insert in Figure 4 for definition of terms.

h denotes the high molecular weight side and l denotes the low molecular weight side of the chromatogram.

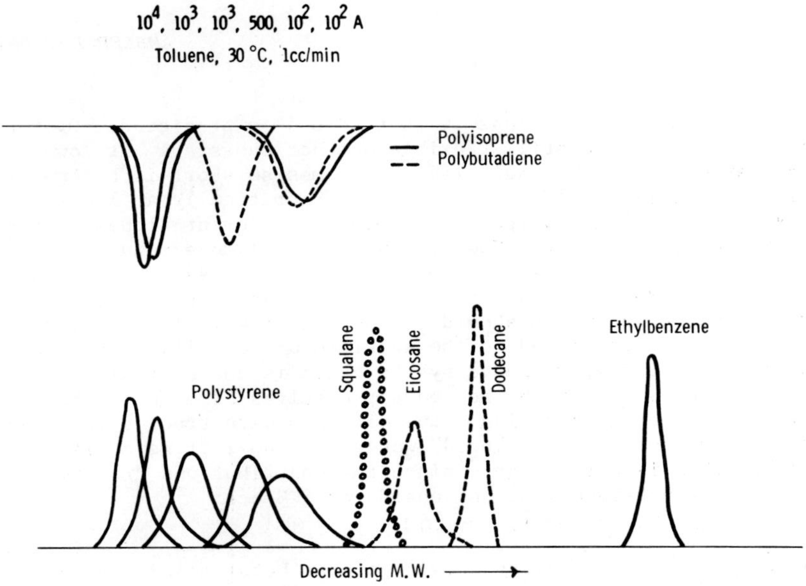

Figure 1. GPC Chromatograms of selected samples.

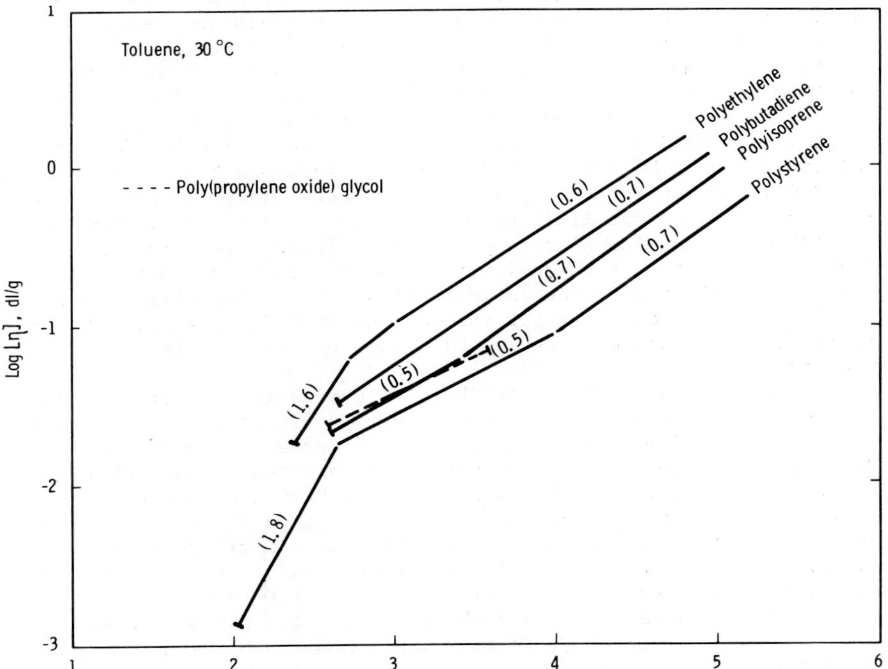

Figure 2. [η]-M Relationships for several polymer types.

motion, a break in the slope to that of 0.5 is predicted. Over this region, the chain is still flexible but not Gaussian. At lower molecular weights, the chain length becomes so short it becomes rigid and the slope increases to 1.8. Theoretically at least, over an extended molecular weight range, three different slopes in the $[\eta]$ -M plot could occur. However, not all polymers behave this way. Various reasons for this behavior have been discussed by Bianchi and Peterlin (Ref. 5) and need not be repeated here. There is, however, one point which should be re-emphasized. In theories of intrinsic viscosity based on the necklace or worm-like chain model, the enhanced viscosity caused by the chain is due entirely to the number of beads, although the beads themselves offer no contribution to the enchanced viscosity. If this were true, a molecule with a length of only one "bead" would have no intrinsic viscosity at all. This is not correct, since the contributions from both the bead and the necklace must be considered. Hence:

$$[\eta] = [\eta]_B + [\eta]_C \tag{1}$$

where $[\eta]_B$ is the intrinsic viscosity of the bead and $[\eta]_C$ is that of the chain. $[\eta]_B$ can be either negative, positive or zero. As polymerization increases the size of the chain, the contributions of $[\eta]_B$ relative to $[\eta]$ become smaller because of hydrodynamic interactions. For example, ethylbenzene, which is the bead in a PSty chain, has a measurable intrinsic viscosity. Thus a correction similar to that of equation (1) is used to explain why certain $[\eta]$ -M plots don't show all three of the predicted slopes.

For PSty, the break in slope from 0.7 to 0.5 occurs at about 10,000 molecular weight. In terms of the GPC universal calibration, this means that for PSty oligomers of less than 10,000 molecular weight, calibration of the GPC cannot be based on the hydrodynamic volume theory of $M[\eta]$. A similar change in slope occurs for PIso and PPOG, but not for PE and PBd. For these samples as well, there is a molecular weight limit below which the samples do not qualify for the hydrodynamic volume calibration. These lines are actual experimental data for the polymer types used in this study. Since there are obvious differences in their solution properties, they should make good examples for the subsequent GPC analyses, particularly when it comes to checking $M[\eta]$ as a universal calibration parameter.

The GPC molecular weight calibration curves (log M versus PEV) for these polymer types are shown in Figure 3. It is apparent that these oligomers had different calibration curves. The values of $M[\eta]$ for each sample were calculated and plotted versus PEV in Figure 4. In this case all samples except PPOG appeared to fit a common curve, with the PPOG samples remaining on the column longer than expected. Dawkins (Refs. 7-9) observed a similar behavior during his GPC experiments with polystyrene under theta conditions, and it was proposed that specific solute gel interaction such as adsorption was a factor in his GPC results. The extreme skewing of the chromatograms of the PPOG samples (see Table II) may also be attributable to absorption. Therefore, two futher experiments were done with PPOG sample 41983.

In the first experiment, it was found that PEV increased as the concentration decreased (top of Figure 5). This is the reverse trend to the usually observed concentration effect in GPC, but the result is that to be expected if adsorption is occurring. If this is the case, an extrapolation to "infinite concentration" should effectively eliminate any extraneous adsorption effects on the PEV by flooding out the adsorption capacity of the column. Interestingly when plotted in this fashion it was found that the data described fairly well a line having a y-intercept (representing infinite concentration) coinciding with that needed to allow this sample ($M[\eta]$ = 279) to fit the universal $M[\eta]$ line of Figure 4.

In the second experiment, the amount of material adsorbed was evaluated as a function of the amount injected. In the bottom of Figure 5, ratio of the chromatogram area of eluted material to the

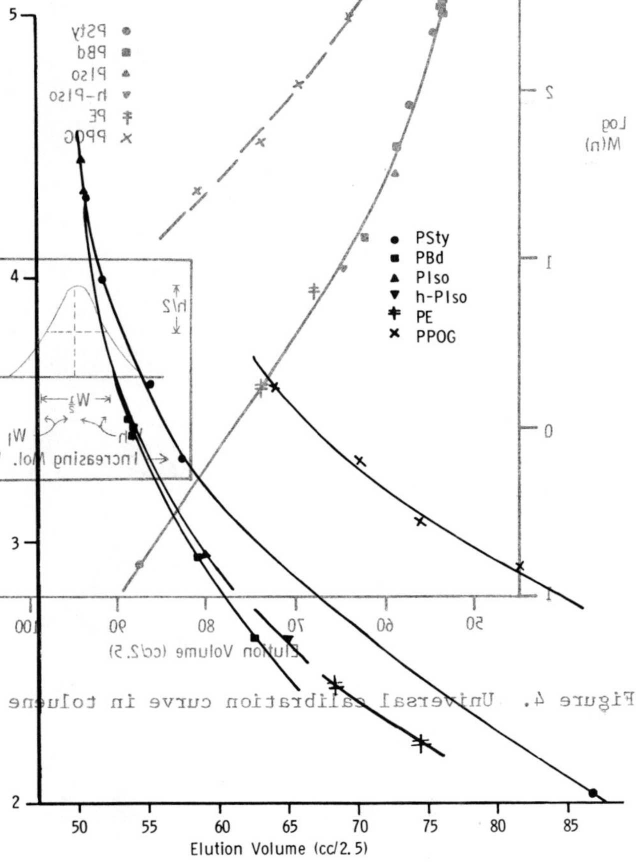

Figure 3. Molecular weight calibration curves in toluene at 30°C.

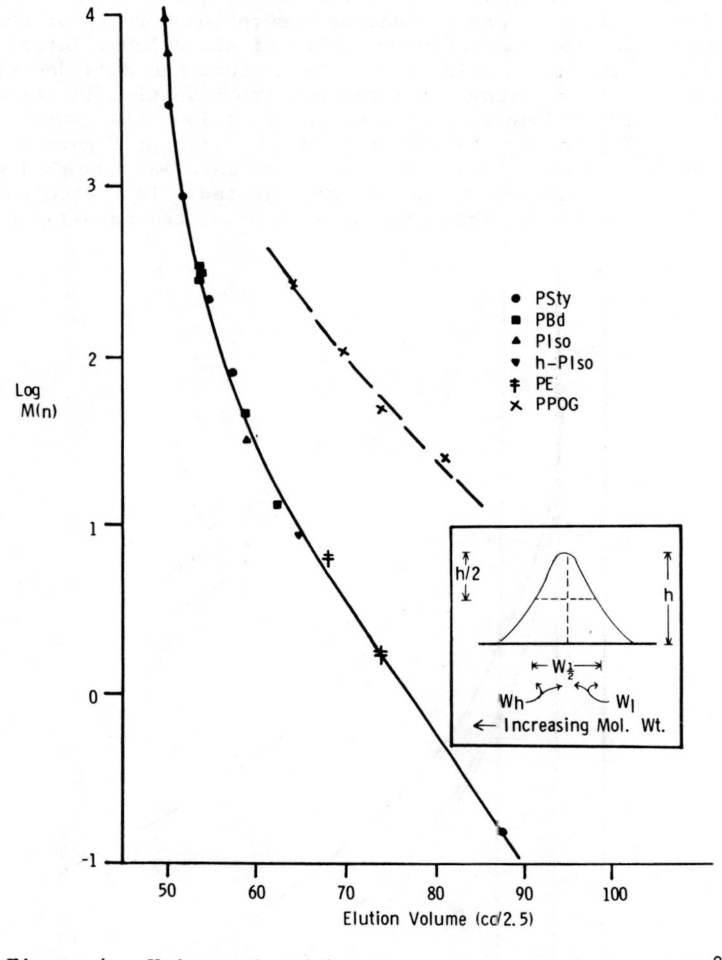

Figure 4. Universal calibration curve in toluene at 30°C.

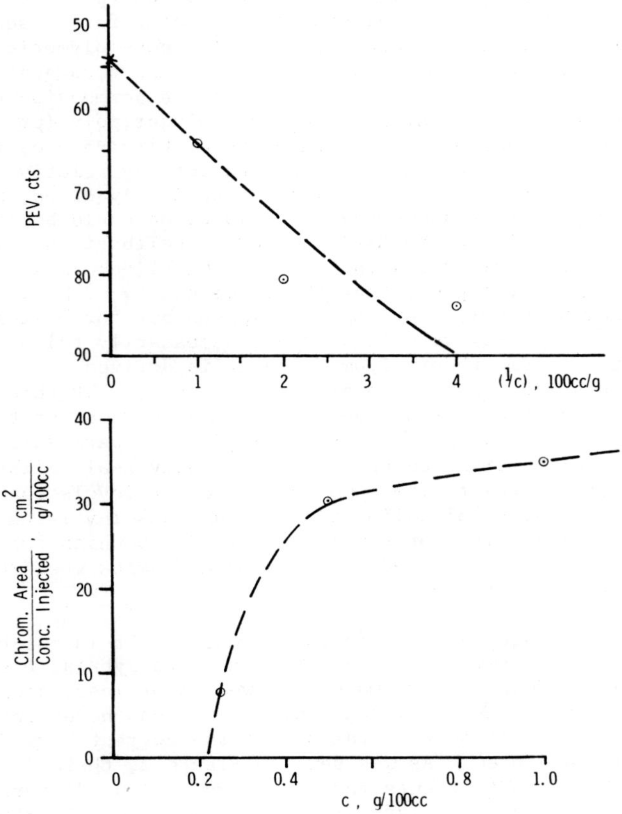

Figure 5. GPC Characteristics of PPOG in toluene at 30°C.

concentration injected was plotted versus concentration. The ratio was found to decrease, suggesting a relative increase in adsorbed material, with decreasing concentration. At concentrations below 0.2 g/100 cc, total adsorption could be expected.

DISCUSSION

Figure 4 indicated that in the absence of adsorption, M[η] functioned as a universal calibration parameter for these oligomers and it appeared to be sustained down to the non-polymeric molecular weight range of 100. The data of Figures 4 and 5 suggested adsorption of PPOG on the column in addition to the permeation process. Under these conditions using toluene as solvent, poly (propylene oxide) glycol cannot be used to generate a universal calibration curve. Presumably, this could hold true for any relatively non-polar solvent. Even THF, if it were scrupulously dried, may show this tendency. It is shown that if adsorption could be eliminated, PPOG also would follow the M[η] universal calibration. The results shown here, when combined with those of the oligomers studied by Belenkii (Ref. 4), suggest that the universality of M[η] exists not only for high molecular weight polymers but for oligomeric species as well. However, it does not necessarily follow that the hydrodynamic volume theory from which it is derived for high molecular weights also applies to oligomers. Indeed, at this time we must admit that there is no theoretical explanation for this result.

From an analytical point of view, it appears permissible to extend molecular weight calculations via "universal calibration" approaches (Ref. 10) to lower molecular weight levels than thought previously. "Universal calibration" techniques involving M[η] appear to be justified for samples of broad MWD which contain oligomeric species, provided suitable Mark-Houwink coefficients for the oligomers are known.

This last point deserves one further comment. It is not uncommon for commercial polymers like polyethylene or polybutadiene to contain oligomers down to 1000 molecular weight or less, and although it has been shown that universal calibration can be extended down to this region, care must be taken that the correct Mark-Houwink coefficients are used. As was seen in Figure 2, there is a good chance that the coefficients for oligomers will be different from these for the high molecular weight region. When calibrating the GPC with polystyrene standards, it must be known with certainty whether or not the Mark-Houwink coefficients change and the approximate molecular weight at which the change occurs before accurate values of M[η] can be calculated. This is an important requirement to the generation of universal calibration data with polystyrene, as well as accurately calculating molecular weight and molecular weight distribution data of uncharacterized polymers.

The authors wish to thank The Goodyear Tire & Rubber Company for permission to publish this work. The polybutadiene and polyisoprene samples were synthesized by S.L. Church.

References

1. Z.Grubisic, P.Rempp and H.Benoit, J.Polym. Sci., B, 5, 753 (1967).
2. M.R.Ambler and D. McIntyre, J. Polym. Sci., Polymer Letters, 13, 589 (1975).
3. W.B. Smith and A. Kollmansbérger, J. Phys. Chem., 69, 4157 (1965).
4. B.G. Belenkii, I.A. Vakhtina and O.G. Tarakanov, Vysokomol. Soedin., Series B, 16, 507 (1974).
5. U. Bianchi and A. Peterlin, J. Polym. Sci., A2, 6, 1759 (1968).
6. W.H. Beattie and C. Booth, J. Appl. Polym. Sci., 7, 507 (1963).
7. J.V. Dawkins, Makromol. Chem., 176 (6), 1777 (1975).
8. J.V. Dawkins, Makromol, Chem., 176 (6), 1795 (1975).
9. J.V. Dawkins, Makromol. Chem., 176 (6), 1815 (1975).
10. M.R. Ambler, J. Polym. Sci., Polym. Chem., 11. 191 (1973).

CALIBRATION AND DATA PROCESSING IN HIGH SPEED GEL PERMEATION CHROMATOGRAPHY

E. Kohn and R. W. Ashcraft

Development Division
Mason and Hanger-Silas Mason Co., Inc.
Pantex Plant
Amarillo, Texas

SUMMARY

When microgel columns are employed in GPC the use of the conventional siphon counter for monitoring the solvent flow is not applicable because of the small amount of solvent delivered during the run. An internal standard method has been investigated which corrects for total change in flow from run to run and permits application of a simple calibration curve to the data. Factors dealing with the choice and concentration of internal standards are discussed. For polymer analyses in THF the peroxide of the solvent was found to be particularly useful as an internal standard, providing the concentration of the peroxide was controlled. Temperature changes in the columns are not corrected by the method.

A Wang 700B desk calculator was used for processing the GPC data, with or without the use of a Grant comparator. When the Grant was used, the strip chart recordings of the detector output were photographically reduced and read on the Grant, which was interfaced with the Wang to yield molecular weight parameters on a print-out. Alternately, the strip chart recordings were visually read and the data were recorded on the Wang which yielded $\overline{M}_n$ and $\overline{M}_w$. A method is also described in which the calculator is directly interfaced with the detector to yield the molecular weight parameters.

INTRODUCTION

Gel Permeation Chromatography (GPC), a technique for the separation of mixtures of molecules according to their sizes, has found extensive use in the analysis of plastics and other materials of commercial importance and is rapidly becoming the technique of choice for the molecular characterization of polymers. Since it is not an absolute method, however, it requires the relating of the retention volume (V_r) of a particular component to its molecular weight by an absolute method such as membrane osmometry and/or light scattering photometry. In many cases the chromatograms are evaluated by relating V_r of components of the sample to a calibration curve which is developed from standards having relatively narrow molecular weight distributions and which have been well characterized by one or more of the absolute methods. In all cases the exact evaluation of V_r is critical since small variations in it result in large changes in the corresponding molecular weight. In practice this evaluation is often performed by a siphon counter which periodically marks the delivery of a measured amount of liquid. But in spite of this procedure significant errors in V_r may be encountered(1).

The recent commercial availability of columns packed with microparticles, such as μ-Styragel*, has facilitated the molecular characterization of polymers because of a five to six fold reduction in V_r over conventional columns, at comparable resolution, and the corresponding reduction in analysis time. In this high speed gel permeation chromatography (HSGPC) V_r's are several tens of milliliters of solvent, as compared to hundreds in conventional GPC, and the errors in small siphon counters which need be used to monitor such volumes become highly significant, generally making such devices inapplicable. Dependence on steady solvent flow rates, even with the best available pumps, is also not satisfactory(2).

---

* *Registered Trademark, Waters Associates*

Several techniques have been described in the literature which are designed to alleviate the flow calibration problem in HSGPC. Bly et al.(*3*) have described a method in which the total V_r is measured by a large siphon counter and V_r's of individual components are calculated by its use. Williams et al.(*4*) have suggested the use of calibration standards which are mixed with the sample in sufficiently low concentration to be undetectable on the primary detector, but observed on a second, more sensitive detector. Patel(*1,5*) has proposed the use of an oligomer of the material under analysis as internal standard for normalization of V_r's. Jones and Runyon(*6*) have described a technique which involves the separate but simultaneous injection of an internal standard with the sample, with the inclusion of a precolumn to delay the emergence of the internal standard peak past the last sample-derived peak. While apparently more or less effective these methods demand the availability of special equipment or require standard samples which are usually not available.

It was found that a low molecular weight substance may serve as an effective internal standard, permitting correction of V_r's which are inaccurate because of overall fluctuations in the flow rate, providing care is taken in the choice and concentration of the standard, and the temperature of the columns is not greatly changed. Since the application of the internal standard requires modification of the usual data handling process this modification was incorporated into several variations of data processing procedures which do not require a dedicated computer.

RESULTS AND DISCUSSION

Flow variations are the principal cause of error in HSGPC analysis carried out in the absence of siphon counters. These errors are primarily of two types, those which have a time scale of about a half hour or longer, corresponding to the length of a single run, and those with a time scale of seconds or a few minutes and which tend to compensate each other over the time span of the run. There is presently no simple method to correct for errors by the latter cause. Fortunately, errors from this source are usually small and if they are significant experimental conditions can often be adjusted to reduce them.

The other type of error, which is caused by gross changes in flow rate from run to run and particularly flow rate changes between calibration runs and sample runs, can be corrected in a number of ways described previously(*1*,*3*,*5*,*6*). It was also found to be correctable by a simple internal standard (IS) technique which consists of normalizing V_r values of the components of the chromatogram to the V_r value of an internal standard which has been previously calibrated. This procedure corrects precisely for changes in the flow rate from run to run but does not correct for fluctuations during the run which do not affect the overall flow rate.

Four relatively low molecular weight substances were evaluated with regard to their suitability as IS's in HSGPC. These were o-dichlorobenzene (ODB), air, cyclotetramethylene tetranitramine (HMX), and the peroxide of tetrahydrofuran (THFP). The first two are small nonpolar molecules which do not hydrogen bond, while the last two are more polar and subject to hydrogen bonding. They were each investigated with respect to repeatability of V_r in different sample environments, and ODB as well as HMX were also examined with regard to the effect of their concentration on V_r. The compounds were evaluated by multiple determinations in the presence of each other or of a polystyrene standard. The results were found to be reproducible and consistent, commensurate with errors inherent in the reading of V_r's, the recorder fidelity, and small local fluctuations. They are summarized in Table I.

Many small molecules are suitable as internal standards but the choice of a good IS must, among others, include the following

Table I. Evaluation of Internal Standards[a]

Substance	Reference	Substance V_r[b]	Deviation (%)[c]	Number of Runs	Reference V_r[b]
HMX	THFP	56.17	0.06	6	59.82
	PS[d]	56.18	0.09	4	42.55
THFP	HMX	59.82	0.14	19	56.18
	PS[d]	59.82	0.15	6	42.55
ODB	Air	61.68	0.04	9	66.24
Air	HMX	66.24	0.09	6	56.18

[a] *At 30°C with columns described in the experimental section. Concentration of HMX 0.05% (w/v), of ODB 0.125 (w/v), others not known but small. UV detector used.*

[b] *Retention volume in mℓ, each recorder chart division was approximately 1.15 mℓ.*

[c] *Standard Deviation.*

[d] *Polystyrene standard, peak mol. wt. 20,000.*

factors: (1) good detector sensitivity, since very small concentrations of the IS are preferred to avoid column loading, (2) the V_r of the IS should be significantly larger than that of any of the sample components; and (3) there should be no molecular interaction with the sample components, such as chemical reaction, hydrogen bonding, or other forms of complexing.

o-Dichlorobenzene and air meet all the above requirements but HMX and THFP do not meet all of them since these compounds are polar and the peroxide, particularly, is fairly reactive. Nonetheless, the latter two were found suitable as standards in the characterization of a number of polymer types. The peroxide is especially convenient when THF is used both as carrier and as sample solvent, which is often the case. In such instances small amounts of the peroxide are always in the sample because of the ever presence of oxygen. Since the peroxide is a strong UV absorber it exhibits excellent sensitivity with this detector and its high refractive index extends its usefulness to the RI detector.

The air peak deserves special note. Air is often inadvertently introduced into the columns as a minute bubble in the injecting syringe or the injector. The tiny air bubble probably dissolves or is absorbed when the carrier liquid enters the columns and is reformed when the liquid emerges from the columns. The bubble produces a prominent, usually positive, peak on the UV detector trace when the instrument performs at high sensitivity and a much smaller negative peak with the RI detector. Because of its low sensitivity it was not used with the latter detector.

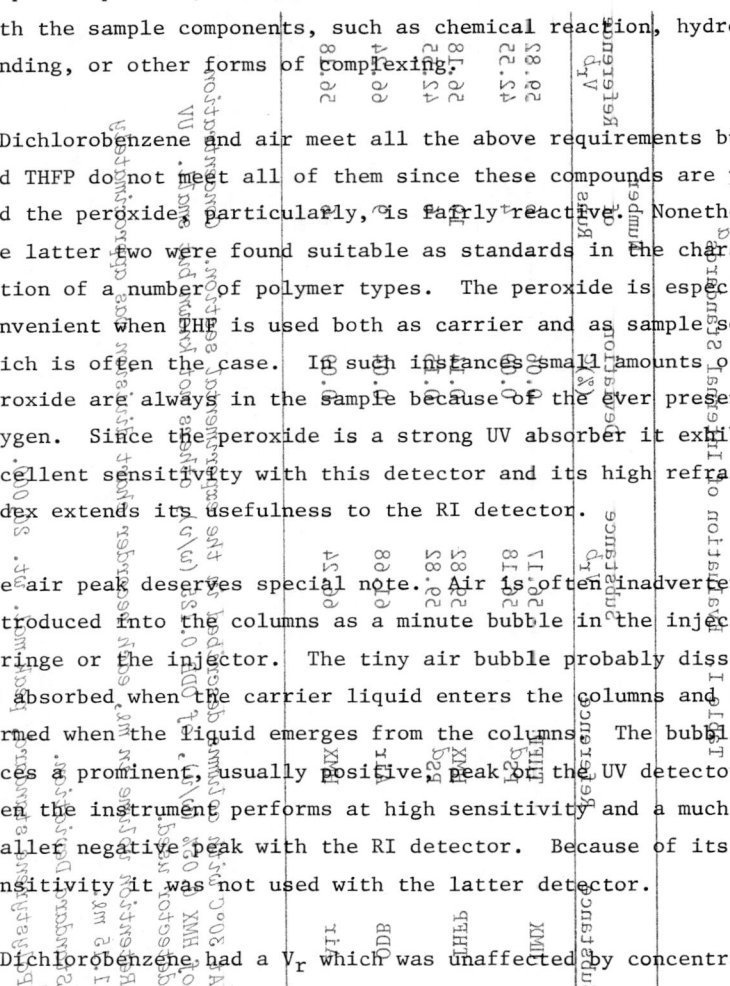

o-Dichlorobenzene had a V_r which was unaffected by concentration in the range of 0.0078 to 1.000% (standard deviation 0.04%) but HMX showed a dependence of V_r with concentration over the range of 0.0125 to 0.200%, the latter being close to its solubility limit in THF. This dependence is shown in Fig. 1, where each

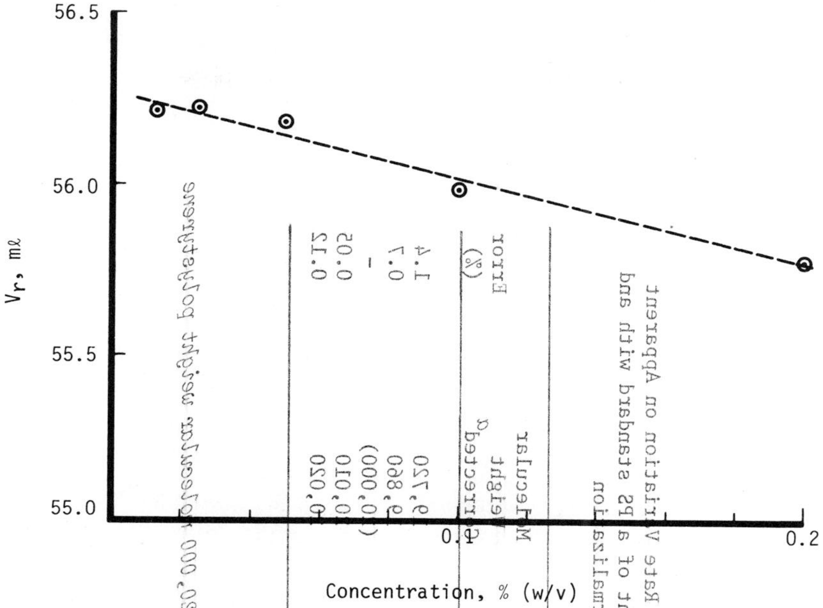

Fig. 1. Effect of concentration on V_r of HMX.

point represents the average of six separate runs. Thus, it is seen that while ODB may be used as an IS without regard to concentration, up to at least 1%, HMX may only be used in this manner if its concentration and corresponding V_r are known.

The utility of the IS method is illustrated by the data in Table II which show the errors in molecular weight of a polystyrene standard that may be encountered as a result of flow fluctuations from run to run and the correction achieved by means of the IS. The data were obtained by deliberately changing the flow rate by ± 5% and interpolating at ± 2.5%. While flow fluctuations of 5% are not common, variations of several percent are frequently encountered, particularly when runs separated by several days or weeks are compared. It is seen from Table II that molecular weight errors from such variations produce unacceptable data but that use of the IS leads to satisfactory results.

Table II. Effect of Flow Rate Variation on Apparent Molecular Weight of a PS standard with and without IS[a] Normalization

Flow Change (mℓ/min)[b]	Molecular Weight (Calculated)	Error (%)	Molecular Weight Corrected[a]	Error (%)
+0.10	36,080	80	19,720	1.4
+0.05	26,480	32	19,860	0.7
0	(20,000)	-	(20,000)	-
-0.05	13,960	30	20,010	0.05
-0.10	9,910	50	20,020	0.12

[a] *HMX*

[b] *IS was calibrated at 2.30 mℓ/min with 20,000 molecular weight polystyrene standard.*

Variations in the temperature of the columns, however, are not compensated by the IS method and may occasionally produce errors of the magnitude of those produced by flow fluctuations. The data which are summarized in Table III show that for small temperature changes (one degree or less) both uncorrected and corrected errors are tolerable. The data also indicate that the temperature coefficient of V_r is different for different molecules, in accordance with previous findings by Patel(*5*). The data clearly demonstrate the importance of good temperature control of the columns if reproducible results are required.

Application of the IS method complicates the data processing procedure since molecular weight moments cannot be continually accumulated as is done in most computerized GPC programs. If the chromatogram on the recorder chart is used the V_r of the IS can be read first and appropriate corrections can be applied to the recorded retention distances converting them to the appropriate molecular weight related parameter. If the detector signal is processed directly, data from the entire run must either be recorded or stored in a memory device to permit application of IS data to it for appropriate normalization of V_r's.

Three techniques were developed in the present study to provide several levels of data processing capability. In the simplest case the detector output was traced on graduated recorder chart paper and was then visually read and the data transferred to a programmable calculator which was appropriately programmed. High data gathering resolution is not required in this procedure since peaks can always be read wherever they occur, including the IS peak, and only about 30 readings are required to obtain reproducibility to 1%. The time for reading is about 20 minutes and one run can be processed while the next one is being recorded. Molecular weight parameters are immediately available and calibration checks can be made as required, along with appropriate small adjustments to the calibration curve.

Table III. Effect of Temperature Variations in Columns on IS Corrected Molecular Weights

Deviation From 30°C	Uncorrected Molecular Weight	Error (%)	Corrected Molecular Weight (HMX)[a]	Error (%)	Corrected Molecular Weight (Air)[a]	Error (%)
10	22,480	12	18,290	9	17,760	11
4	21,100	6	19,390	3	19,160	4
2	20,570	3	19,700	1	19,600	2
1	20,280	1	19,900	0.5	19,780	1
0	(20,000)[b]	-	(20,000)[b]	-	(20,000)[b]	-
1	19,750	1	20,250	1	20,200	1
2	19,450	3	20,430	2	20,530	3
4	18,880	6	20,860	4	21,080	5
-10	17,000	15	22,000	10	23,450	17

[a] IS

[b] *Either HMX or air was calibrated against 20,000 molecular weight polystyrene standard. Calibration curve used was that described in the experimental section.*

For more routine and exact data processing the recorder trace was read with a comparator which was interfaced with a programmable calculator. Double the data gathering resolution was employed in this technique with about 60 readings per run. Check runs showed that greater resolution did not improve the precision of the results.

In the third method the detector was interfaced with the programmable calculator, the data were stored in memory and the internal standard was visually read after the completion of the run and entered in the calculator. Because of the relatively small number of storage registers the data resolution was limited but this could have been corrected by addition of more storage capability. The necessity of visual reading of the internal standard prevents this method from being completely automated but its elimination appears to be beyond the capability of presently available calculators since it would require a program which could evaluate the IS peak with a resolution of about 50 microliters in V_r, with a comensurately large storage and programming capacity.

EXPERIMENTAL

A Waters ALC-100 liquid chromatograph was equipped with a 6000-A pump, a U6K high pressure injector, a R-400 differential refractometer and a 440 absorbance detector, operating at 254 nm. A three-solvent manifold was also used as part of the solvent train.

Six μ-Styragel columns, equipped with constant temperature blocks, were used in series. They had nominal pore sizes of 100 Å, 500 Å, 10^3 Å, 10^4 Å, 10^5 Å, and 10^6 Å and were arranged in that order, with the sample entering the smallest pore size column first. A 2095 Forma Scientific Co. constant temperature bath, equipped with dual pump, was used to circulate water through the blocks. The bath temperature was controlled to ± 0.05°C and the temperature of the blocks varied not more than 0.05°C from that of the bath.

The output from the detectors was graphed on a Perkin-Elmer 1 millivolt dual pen recorder which was normally operated at a sensitivity of 5 millivolts and a chart speed of 5 millimeters per minute. Chart paper divisions were 0.1 inch.

The solvent used was spectroscopic grade tetrahydrofuran (THF), supplied by Burdick and Jackson Laboratories. It was unstabilized, distilled in glass and stored over nitrogen, with a UV cutoff of less than 212 nm and a maximum water content of 0.03%. The solvent was deaerated prior to use by means of vigorous magnetic stirring under partial vacuum until bubble evolution was not detectable for a period of five minutes. The solvent was then immediately transferred to the chromatograph and placed under a blanket of pure nitrogen. The flow rate was in the range of 2 mℓ per minute at an inlet pressure of about 2,000 psi.

Sample solutions were prepared on a weight per volume basis (w/v), were clarified by passage through a one micron Millipore* filter and stored in bottles capped by Teflon† Mininert valves. Normally, 50 μℓ of the solution was introduced into the loop of the injector, the lever was flipped and the contents of the loop instantaneously became part of the solvent stream of the column. The event was registered on the recorder trace by a blip which exactly defined the injection point. In experiments where the air peak was used as IS and the UV detector was operated at a low sensitivity a minute air bubble was placed in the front end of the syringe barrel.

The recorder tracings of the detector were either read with a calibrated loupe, a Grant comparator, or directly processed by a programmable Wang 700B calculator. In the first instance, injection points, reference peak, and sample peaks or points along the

Registered Trademark, Millipore Corporation

†*Registered Trademark, DuPont Corporation*

distribution curve were read as chart-paper divisions, to the nearest hundredth of a division. Points along the molecular weight distribution curve were similarly read with respect to height from a baseline previously determined. The values were manually entered into the calculator and processed according to a program which converted the raw retention distances on the chart to corrected distances by means of the relation:

$$D_{corr} = D_{obs} \times \frac{IS_{cal}}{IS_{obs}}$$

where D_{corr} is the corrected retention distance and D_{obs} is the observed retention distance of the peak or position on the curve, and IS_{cal} and IS_{obs} are the calibrated and observed retention distances of the peak of the IS, respectively. The program then converted the retention distance to extended chain length by means of a five parameter fit of the calibration data, and in turn to molecular weight by use of the appropriate Q-factor (41.4 for polystyrene). The molecular parameters such as $\overline{M}_n$, and $\overline{M}_w$ were then computed by the Wang and a distribution curve was plotted by a plotter-typewriter, if so required.

In the second instance the chromatogram charts were photoreduced with a copy lens having low linear distortion. The negatives were then read on a Grant comparator which was interfaced with the Wang for direct processing of the readings and development of the molecular weight parameters.

In the most direct data processing mode the detector output was interfaced with the Wang by means of a Fluidyne Instrumentation system which consisted of a 7110A autoranging digital multimeter, a 728-700 digital interface, a 7220A digital timer and accessory equipment. The injection point was recorded by starting the program and after a scheduled waiting period, covering the elution of pure solvent, times and amplitudes of the chromatogram were stored in 15 second intervals. The peak of the IS was manually marked and entered into the program which then produced the molecular weight parameters.

The calibration curve was developed by use of 15 narrow distribution polystyrene standards ranging in molecular weight from 1400 to 2,300,000, which were obtained from Pressure Chemical Co., and Waters Associates, and six straight chain hydrocarbons with 10 to 36 carbon atoms, obtained from Chemical Services, Inc. Retention distances of peaks of these materials, corrected by means of the IS, were then plotted on semilogarithmic graph paper with the retention distances on the linear axis and the molecular weights on the log axis. A smooth line was drawn through the

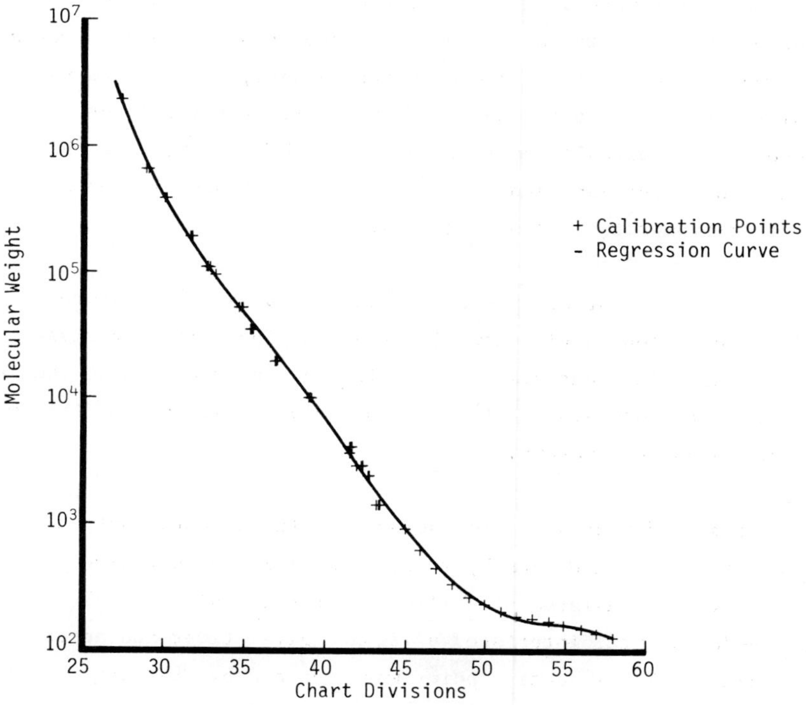

Fig. 2. Polystyrene molecular weight calibration data and regression curve.

polystyrene points and another through the hydrocarbon points. An adjustable French curve was then fitted to the hydrocarbon points and was moved upwards, without changing slope, until it formed part of the polystyrene curve. Points were then taken from the displaced hydrocarbon curve and were made part of the set of data which was used to develop a fifth degree polynominal calibration function through use of a stepwise linear regression program(*7*). Fig. 2 shows the computer derived calibration curve and the experimental points.

CONCLUSIONS

The internal standard method, employing a small more or less polar molecule was found to be effective in reducing errors caused by gross flow variations in high speed gel permeation chromatography. The choice of the standard, and in some cases its concentration are relevant. The method does not correct for major temperature changes in the columns but can be used providing such changes are small. Data processing techniques, requiring relatively unsophisticated equipment, can be adapted to incorporate the internal standard correction except that for completely automatic data processing more extensive memory and programming capabilities are required than is normally available in a programmable desk calculator.

ACKNOWLEDGEMENTS

We thank the Energy Research and Development Administration, under whose auspices this work was performed, for permission to publish. We are also indebted to James E. Matlock for assistance with the experiments.

REFERENCES

1. G. N. Patel and J. Stejny, J. Applied Polymer Sci., 18, 2069 (1974).

2. D. D. Bly, H. J. Stoklosa, J. J. Kirkland, and W. W. Yau, Anal. Chem., 47, 1810 (1975).

3. D. D. Bly, W. W. Yau, and H. J. Stoklosa, Anal. Chem. 48, 1256 (1976).

4. R. C. Williams, J. A. Schmit, and H. L. Suchan, Polymer Letters, 8, 413 (1971).

5. G. N. Patel, J. Applied Polymer Sci., 18, 3537 (1974).

6. J. Jones and J. Runyon, A Device to Provide a Non Interfering Internal Standard in Gel Permeation Chromatography. Presented at the International GPC Symposium, Pittsburgh, Pa. (October, 1975).

7. W. J. Dixon, Ed., Biomedical Computer Programs, University of California Press, Berkeley, CA (1971).

SOLUTION OF MATERIALS PROBLEMS IN FOREST PRODUCTS

J. Cazes and N. Martin

Waters Associates, Inc.
Milford, Massachusetts

INTRODUCTION

The problem involving production of unacceptable products from wood is often caused by the materials - resins or chemicals - that are used in conjunction with the wood. A new technology for analyzing resins and other complex mixtures has shown itself to be a powerful tool in solving these problems. This technology, liquid chromatography (LC), with its related technology, gel permeation chromatography (GPC), has been applied in a number of areas in the wood industry. In the production of particleboard and plywood, for instance, GPC has been used to show up batch-to-batch variations in incoming resins which could lead to disastrous production failures, and to compare and evaluate raw materials from new sources. In post-treatment of wood for outdoor use, liquid chromatography provides a specific method of analysis of the preservative in samples from wood borings. Liquid chromatography can also be used to monitor plant effluent for potential pollutants.

GPC and LC both use the same basic instrumentation. GPC is used primarily for resins, LC for lower molecular weight organic chemicals. Both these methods of analysis are basically the same. Sample preparation is minimal - frequently, solution and filtration are all that is required. Once the method has been developed for a specific application, typical analysis times are less than ten minutes for LC, often as little as 20 minutes for GPC. Some situations where GPC and LC have been used are outlined in the flow chart given in Figure 1.

Let's look at the kind of information that GPC and LC can provide. The resins used in fabricating wood products are complex mixtures of polymer, additives, and impurities. Like other natural and snythetic polymers, they are distributions of molecules of different chain lengths, and no two batches are ever exactly the same. Subtle differences between batches can have marked effects on the

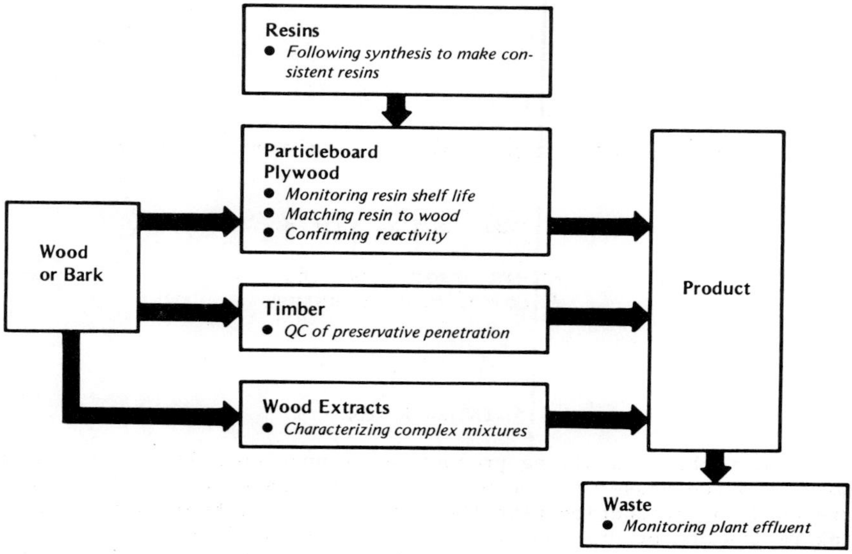

Figure 1. Applications of GPC/LC in the wood industry.

processing and end-use characteristics of the resin. Gel permeation chromatography, GPC, is a technique for separating the resin into its component parts and characterizing it by one of its fundamental properties, the molecular weight distribution. Resin properties that are affected by molecular weight distribution include viscosity, reactivity, cure rates, adhesive tack, bond strength, and many others.

In GPC, a sample of the resin is dissolved and injected into a flowing stream of solvent which passes through a rigid, porous gel whose pore sizes vary over a specified range. Molecules which are large in relation to the pore sizes enter very few pores, and emerge from the column with little delay. Smaller molecules enter more of the pores, travel a much more involved route, and take longer to emerge from the column. Thus, larger molecules emerge from the column first with smaller ones eluting later. As each component emerges from the column, its presence is detected, and shows up on the resulting chromatogram. (Figure 2)

Often, one finds that two batches of a given resin exhibit similar bulk properties such as viscosity, specific gravity, etc., but do not perform identically in a given application. This is sometimes due to the fact that they may have the same molecular weight averages but difference molecular weight distributions. This is illustrated, for a hypothetical pair of resins, in Figure 3. It is possible, for the two resin distributions shown here, to have identical averages but very different distributions.

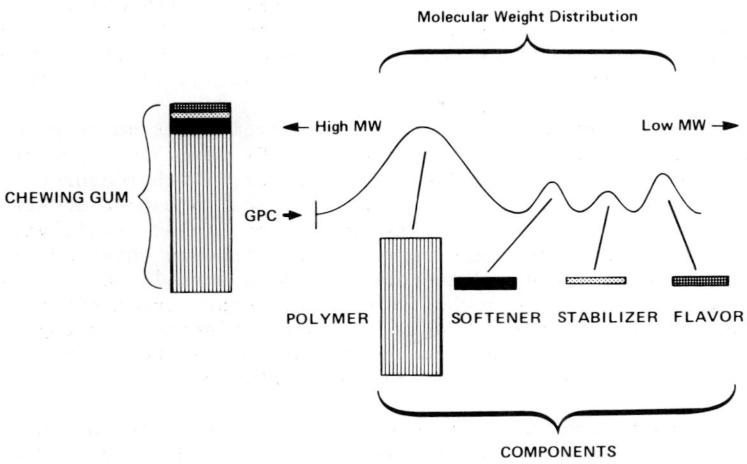

Figure 2. GPC separates mixtures into their high, medium, and low molecular weight components.

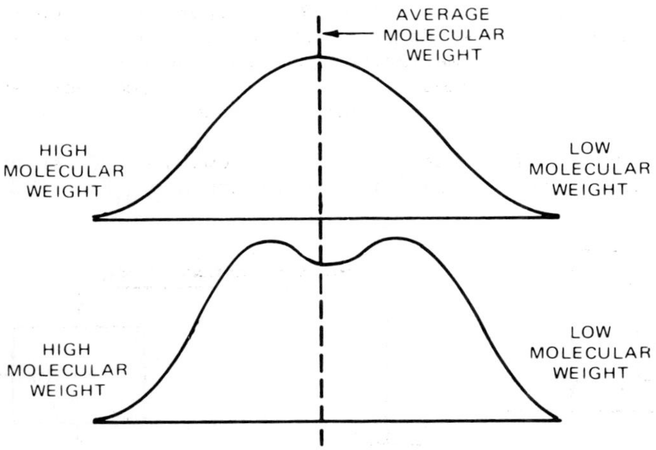

Figure 3. Average molecular weights or bulk properties may not show up significant differences between batches.

A typical system used for performing GPC/LC analyses is shown schematically in Figure 4. The solvent delivery system draws solvent from a reservoir and pushes it through the injector where a solution of the sample is introduced. From the injector, the flowing solvent stream, carrying the sample, passes through the column where the sample is fractionated. The separated sample components then flow through detectors for quantitation and identification.

The detector used to monitor the emerging stream is generally either a differential refractometer or a UV absorbance detector. The refractometer is a "universal" detector, which measures the difference in refractive index (RI) between the component and the solvent stream. The UV detector is capable of higher sensitivity than the refractometer, but only responds to compounds which absorb UV. Either or both of these detectors have been used in the work included here. Electrical signals from the detectors are monitored by a strip-chart recorder which draws a characteristic chromatogram. If desired, sample components can be collected, as they emerge from the instrument, for further analysis and identification.

The following sections describe some interesting and useful applications of GPC/LC to materials problems in the wood products industry.

MONITORING PHENOLIC RESIN SYNTHESIS

A common problem in the production of thermosetting resins involves attempts to make a resin whose properties are consistent from batch-to-batch. The kinds of tests that are normally used to follow the course of the polymerization process show average properties of the resin - pH, acid number, viscosity - which may not reveal subtle, yet significant, differences between batches.

The series of curves in Figure 5 shows how GPC can be used to monitor the course of phenolic resin synthesis. Each GPC run took

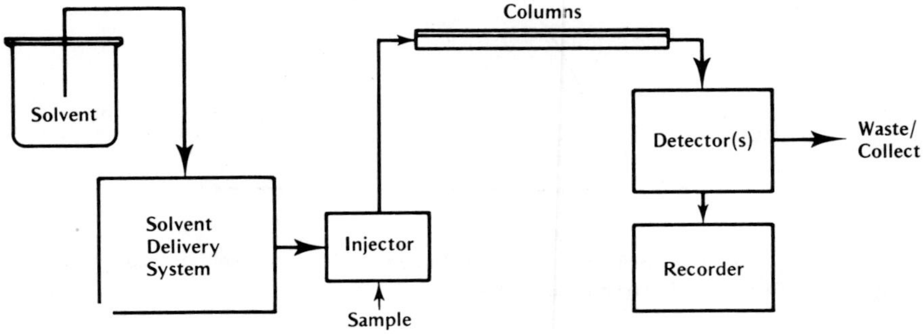

Figure 4. A typical system for GPC and LC.

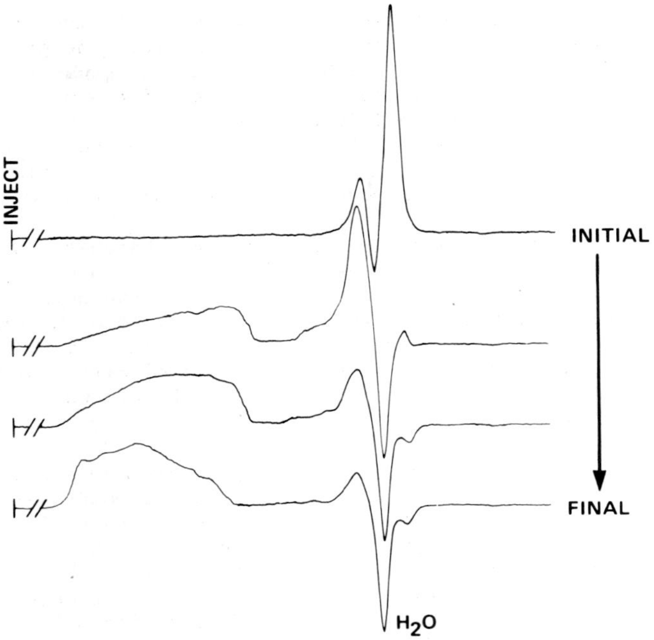

Figure 5. Following the synthesis reaction closely allows you to improve batch-to-batch reproducibility. (Data reproduced by courtesy of Dr. Chung-Hse, Southern Forest Laboratory, USDA, Pineville, L.A.)

10-12 minutes. This was rapid enough that GPC could be used as a production control tool. In such a situation, the polymerization would be stopped when the "fingerprint" fell within the acceptable range of the resin. In Figure 5, initially, only low molecular weight starting materials such as phenol, formaldehyde, and water are present. As the reaction proceeds, high molecular weight material is formed (shown at the left side of the curves). Since the performance of a resin depends upon how much high molecular weight material is present and upon the shape of the overall molecular weight distribution curve, the rapid characterization of a resin via GPC is a powerful tool in producing material with consistently good properties.

TAILORING ADHESIVE RESINS TO APPLICATIONS

In plywood production, it is important that some resin stays on the surface of the wood before lamination rather than completely soaking into the wood. "Dryout" is most likely to occur with a very porous wood, such as southern pine.

The viscosity of the resin depends on the relative proportions of high and low molecular weight materials in it. The greater the proportion of high molecular weight material, the greater will be the viscosity. GPC gives you a "fingerprint" of the molecular weight distribution of a specific batch of resins. The curve in Figure 6 shows that the glue normally used for southern pine contains more material under the high molecular weight portion of the curve than the glue used for Douglas-fir, a less porous wood.

The resins used always vary somewhat from batch-to-batch, and solids content and viscosity measurements made at room temperature do not accurately predict how the resin will behave before cure. Batches of resin with similar relative proportions of high and low molecular weight material have been shown to behave similarly during the production processes. Comparing GPC data of different batches of resin lets you isolate potential problems before they lead to production problems.

Figure 7 compares a phenolic resin glue from an alternate source to the glue normally used by a plywood plant for Douglas-fir. Two things are worth noting in this curve - firstly, there is a bump on the high molecular weight portion of the curve which is due to the presence of a material added by the manufacturer of the alternate resin to increase the tackiness during the lay-up process. Secondly, the PF resin in the alternate glue showed less high molecular weight material than the standard glue. To make it useable, it was "cooked" with more formaldehyde to make the molecular weight distribution similar to that of the standard resin.

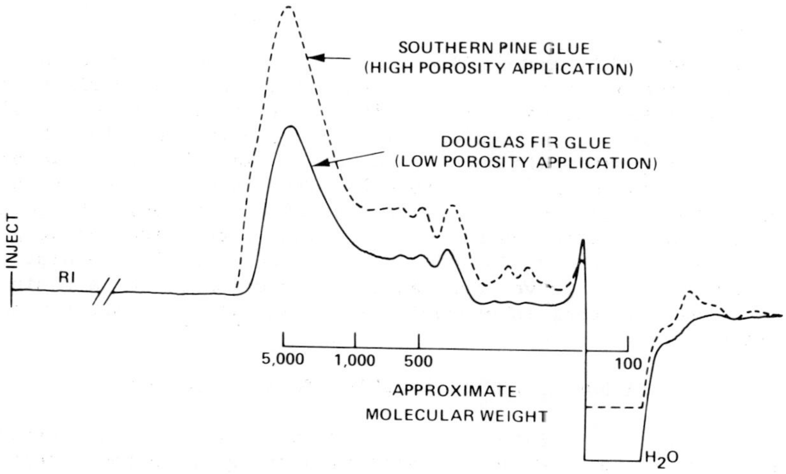

Figure 6. The choice of glue depends on the porosity of the wood.

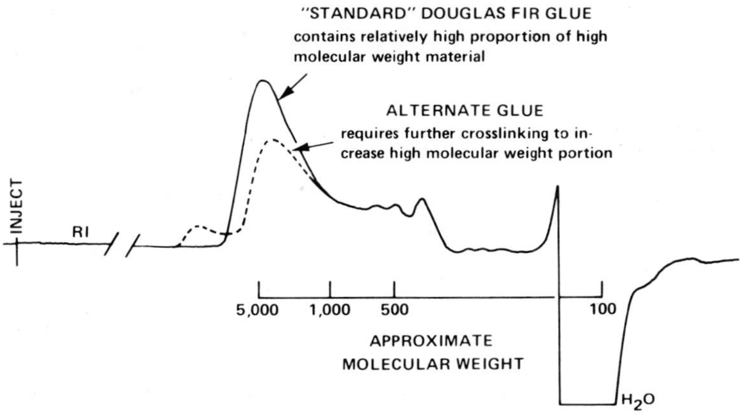

Figure 7. Unsuitable glue can be modified to match the requirements of the application.

PREDICTING RESIN REACTIVITY

To cure satisfactorily, the reactivity of the resin used in particleboard manufacture must be matched to the cure conditions - temperature, pressure and time. The number of available reactive groups varies from batch-to-batch. If the resin is deficient in reactive groups, cure will be incomplete and the particleboard may literally fall apart. The reactive groups in urea formaldehyde resin absorb ultraviolet radiation, and will therefore show up on the UV detector trace of the chromatogram. Lack of absorbance in the UV provides a clear warning that the resin is unreactive and may not cure properly.

Note: All the urea formaldehyde resins analyzed indicated much greater molecular weights than expected - up to 5×10^6. This may indicate the presence of previously unrecognized "giant" molecules which act as large molecules. In either case, the material in solution behaves as if it contains molecules or very high molecular weight.

Figure 8 reveals the differences between a core and face U/F resin. The processing temperature at the face of a thick (1/2" to 1") particleboard is higher than the core temperature, since the face is in contact with the hot press platens. To compensate for this, the resin used for the core must be more reactive than the face resin to assure complete cure at the lower temperatures. The UV traces for commercial core and face resins, shown here, reveal that, while the molecular weight distributions of the two resins are similar, the core resin contains more reactive groups, especially at higher molecular weights.

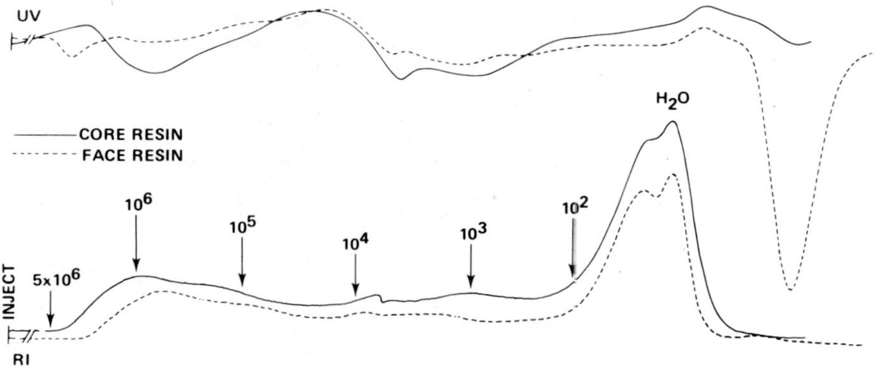

Figure 8. Resins of different reactivity are used in the same board.

The two batches of U/F resin for which the chromatograms in Figure 9 were recorded had identical viscosities, specific gravities, solids levels and pH. However, when the "bad" resin was used, the particleboard fell apart - causing a costly and embarrassing plant shut-down. The UV curve shows why: The "bad" resin contained far fewer reactive groups, leading to very limited cross-linking.

MONITORING RESIN SHELF LIFE

During storage, urea-formaldehyde resins continue to react, increasing in molecular weight and therefore, viscosity, leaving fewer reactive groups to take part in the curing reaction during processing. The molecular weight distribution given by GPC is a very complete characterization of the resin, which can be used to predict viscosity and processing behavior. However, for quality control, it is convenient to compare numbers rather than complex curve shapes - if useful numbers can be easily derived to make a decision whether or not to reject. A simple approach to obtaining numerical values from the GPC curve is shown in Figure 10. Besides the basic application - QC of incoming resins - the same numbers could be used to blend out-of-specification resins to give resins whose viscosity and performance will be satisfactory. A simple mathematical treatment of the curve, shown in the figure can lead to the development of a "go/no go" numerical specification for quality control, and allows batches to be blended to an acceptable viscosity and molecular weight distribution.

Referring to Figure 10, we see that, over a period of time, the proportion of low molecular weight components (indicated by height D) decreases, while high molecular weight components (A and B) increase. The height at C remains constant, indicating that C is probably an intermediate which reacts further to form more A and B. By measuring the heights A, B and D, and finding the relative percentages it is possible to obtain a set of numbers which can be used to specify satisfactory material.

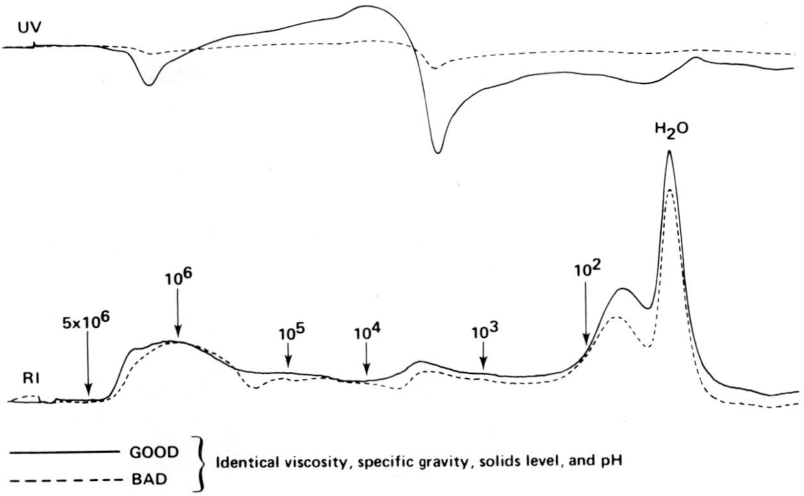

Figure 9. The UV curve shows up unreactive resins.

	HIGH MW COMPONENT		LOW MW COMPONENT	BROOKFIELD VISCOSITY, cp
	% A	% B	% D	
Initially	12.7	12.2	60.7	~175
Final (31 days)	15.8	15.2	54.5	590 (too viscous to process)

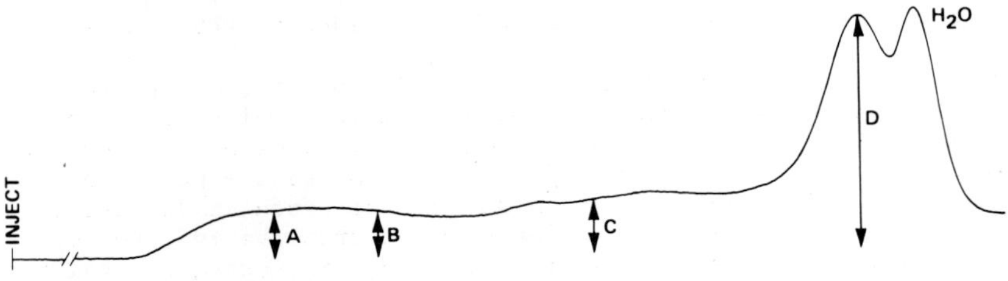

Figure 10. Numerical treatment of GPC curve for monitoring resin shelf life.

QUALITY CONTROL IN WOOD PRESERVATION

Measurement of the level of pentachlorophenol preservative in wood borings is essential in monitoring the effectiveness of the preservative treatment process and meeting buyers specifications. A recently developed LC technique for this measurement has significant advantages over traditional methods which involved ashing, followed by argentometric determination of total chloride. Since total chloride is measured in the traditional approach, it is im-

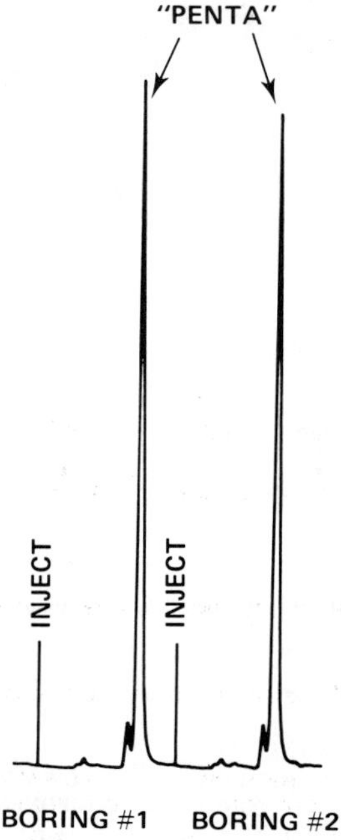

Figure 11. Rapid and specific analysis for pentachlorophenol.

possible to differentiate between penta- and less effective di- or trichlorophenols and other inorganic chlorides in the sample.

The liquid chromatographic method is specific for pentachlorophenol. It requires less time, so that more samples from a batch can be run and a decision on the acceptability of a batch made with greater confidence. It can also be used to measure penta in treatment plant effluents. The concentration of penta at different depths in the wood indicates how long the preservative treatment will last. The levels of penta in acetonitrile extracts from two 3" wood borings are shown in Figure 11. The penta shows up distinctly from other components in the sample. This determination is rapid (less than 8 minutes) and specific.

CHARACTERIZING BARK EXTRACTS

The complexity and variability of wood extracts has limited their use as substitutes for petroleum-based products - even though the price of petroleum products is rising and wood represents a uniquely renewable natural resource. Even natural extracts can, however, be successfully and rapidly analyzed by GPC and LC, so that

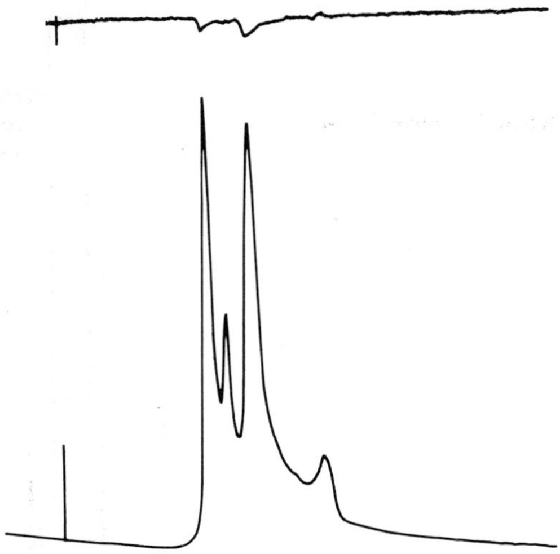

Figure 12. GPC for rapid characterization of bark extracts.

variations in composition and reactivity between batches can be recognized and subsequent processing adjusted accordingly. For example, phenolic extracts from wood bark are now being used as substitutes for petroleum-derived phenols in the production of phenol-formaldehyde resins for particleboard. GPC analysis of the extract shows distinct components in the extract. The relative levels of different components will vary depending on the source of the bark used.

In Figure 12 each peak on the curve for a phenolic bark extract represents a different class of components in the extract. The reactivity of the extract in the production of phenolic resin depends on which components are present in the extract, and in what amounts. GPC gives a rapid characterization of the extract which can lead to better control of the final product.

POLLUTION CONTROL VIA PLANT EFFLUENT ANALYSIS

Growing concern for environmental quality has led to a need for a reliable, reproducible way to monitor trace levels of organic materials in process and plant effluent waters. The first problem is to concentrate the pollutants/contaminants in the sample so that the constituents can be measured and identified. Here, the special properties of one type of LC column packing material are particularly useful. This material has a surface which acts as a better solvent for organics than does the effluent water. If a large volume of the water sample is pumped through the column, the organic compounds in

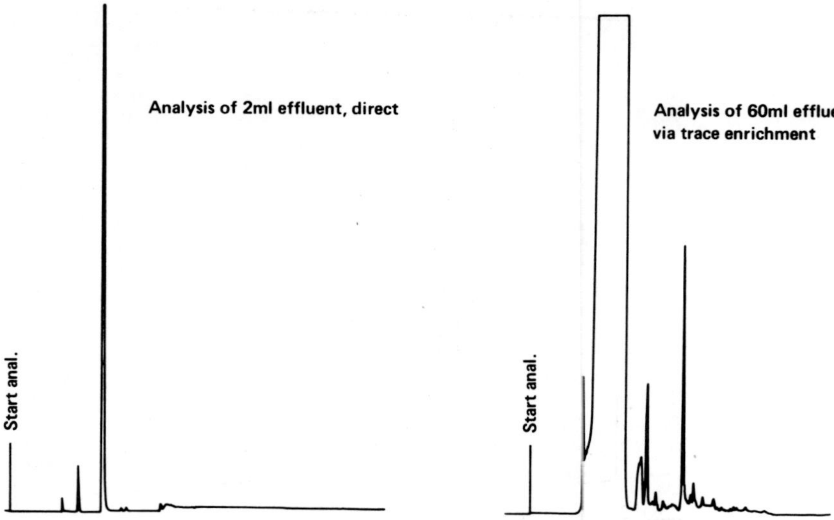

Figure 13. Organic contaminants in plant effluent.

the water have an affinity for the packing material surface, and will stick to the column. When enough sample has been pumped through, then a solvent with an increasing proportion of methanol or acetonitrile mixed with water is pumped across the column. This forces the organics off the column sequentially, with each component emerging as a distinct peak. As the components emerge, they can be collected for further use, or the UV trace can be used as a "profile" of the water. The profile or "fingerprint" will show both number and levels of these components. In this way, changes in water composition during plant use and subsequent treatment can be monitored and the presence or specific compounds tracked.

This technique, called "trace enrichment", was used (Figure 13) to check the level of organic contaminants present in water from a plant source. Direct analysis of the organics in a 2ml water sample revealed the presence of only one major organic component. However, when the organics from 60ml of the water were concentrated directly on the chromatographic column, via trace enrichment, the presence of a number of other components was observed, some of these at parts-per-million levels.

CONCLUSION

We have seen how GPC/LC can be used to solve a broad variety of materials-related problems. The specific applications described here represent only a small portion of the scope of usefulness of this versatile technique. A major advantage resides in the fact that it is a separation method --- materials are separated into their component parts -- and, thus, can reveal subtle, yet significant differences between different batches of a given material. Its sensitivity is limited only by the mode of sample handling employed. The trace enrichment approach, for instance, has been applied to the detection of organic pollutants at the parts-per-trillion level. GPC/LC has been shown to be a valuable tool for the quality assurance, characterization, and production control of both high molecular weight resins and low molecular weight organic chemicals. It is rapidly becoming the method of choice for the solution of materials problems in the forest products industry.

SOLUTION AGGREGATION AND MOLECULAR WEIGHT IN POLY(VINYLCHLORIDE)

R. P. Chartoff
S. K. T. Lo

Department of Chemical and Nuclear Engineering
University of Cincinnati
Cincinatti, Ohio

I. INTRODUCTION

In previous papers (1-3) we reported observations of anomalies during precipitation fractionation of PVC from THF solution, with H_2O as non-solvent. Our data, obtained using GPC and solution infrared techniques, indicate that separation during the fractionation takes place by both stereoregularity and molecular weight because of the formation of molecular aggregates during the precipitation and collection of fractions. The separation by stereoregularity dominates at the beginning of the fractionation with formation of aggregates which are thought to be comprised of highly syndiotactic molecular segments. These tend to precipitate before segments of lower stereoregularity but higher molecular weight.

In GPC the presence of aggregates produces a second, high molecular weight peak on the GPC curve. In some cases the aggregate peak appears innocently as a tail or shoulder at the large hydrodynamic volume end of the GPC chromatogram, while in others it is quite distinct giving the GPC curve the appearance of a bimodal distribution. These different cases are illustrated by curves A (3) in Figures 1-3. The discrete nature of the aggregate portion when it is not initially distinct, may be demonstrated by GPC characterization in the recycle mode. Starting with a

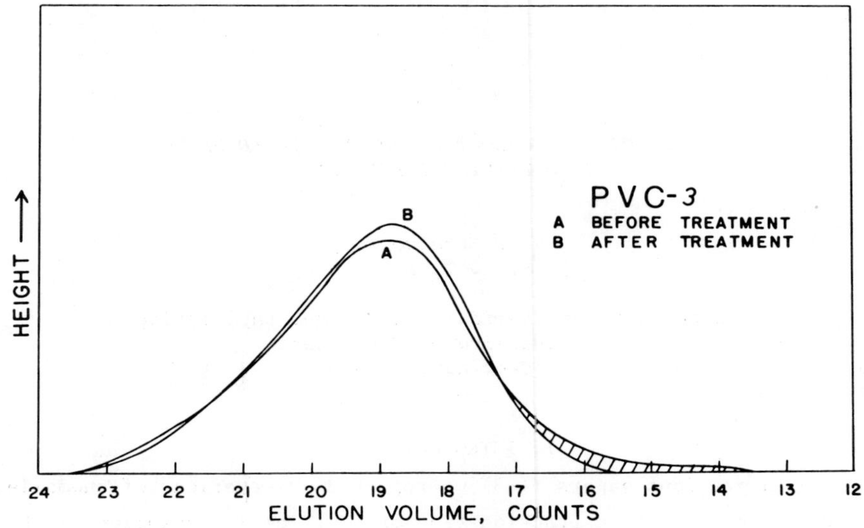

Figure 1: GPC chromatograms for PVC-3 whole polymer; the effect of dissolving aggregates is to convert curve A into curve B (reference 3).

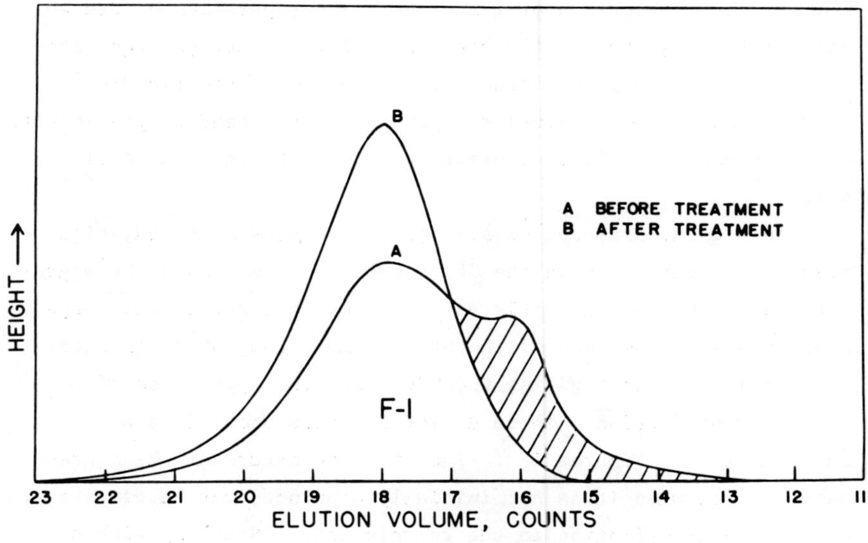

Figure 2: GPC chromatograms for PVC-3 fraction 1; aggregate peak forms a shoulder on GPC curve A and disappears after heating (reference 3).

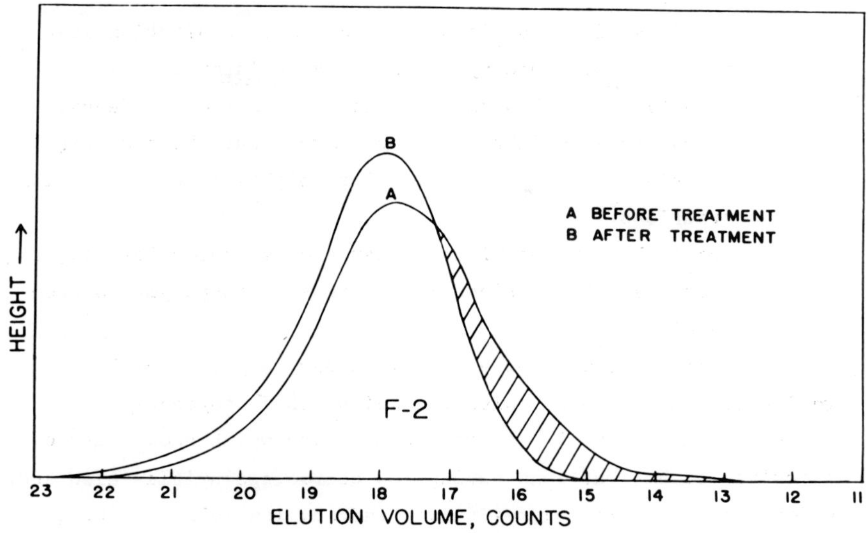

Figure 3: GPC chromatograms for PVC-3 Fraction 2.

fraction having an aggregate shoulder similar to that in Figure 3 we are able to achieve complete resolution of the aggregate peak from the soluble portion on the third pass through the GPC column set.

Generally the aggregates disintegrate into single molecules when heated and the high molecular weight GPC peak disappears as shown in Figures 1-3 by curves B. Unless care is taken to disperse the aggregates by appropriate heating of the solutions used for GPC characterizations molecular weights recorded for the first few fractions may be seriously overestimated.

The thermal conditions necessary for breaking up the aggregates depend on the particular PVC sample of interest (3-7). In some cases, however, where the aggregation tendency is particularly great it is difficult to disperse the aggregates. Some of the problems we have encountered in this regard are discussed subsequently in another section.

Several additional observations support the prevailing view (4,5,7) that aggregation in PVC is directly related to stereoregularity. These include the following:

1) The first few fractions obtained in precipitation fractionation are more syndiotactic than later fractions (2,3,4) these show the greatest aggregation tendency.
2) A "molecular weight inversion" may occur where successive fractions have higher molecular weights than the initial fractions.
3) Among PVC samples with similar average molecular weights anomalies in fractionation increase with polymer stereoregularity.

While the tendency for aggregation may be great for highly syndiotactic and/or high molecular weight chain segments, the purpose of this paper is to present new data which show that low molecular weight fractions having low stereoregularities also show evidence of molecular aggregation. Indeed, low molecular weight PVC has a greater aggregation tendency than any of the other samples we have yet studied.

II. EXPERIMENTAL

A. MATERIALS

PVC resins used in this study were supplied by B.F. Goodrich Chemical Company. Experimental data for three samples are discussed:

PVC-2 having $\overline{M}w$ = 73,900 and $\overline{M}w/\overline{M}n$ = 2.39,
PVC-3 with $\overline{M}w$ = 86,000 and $\overline{M}w/\overline{M}n$ = 2.80,
and PVC-5 having $\overline{M}w$ = 30,600 and $\overline{M}w/\overline{M}n$ = 2.56.

All three were polymerized in suspension, PVC's - 2 and 5 at 50°C and PVC-3 at -15°C. PVC-3 is more syndiotactic than the others (3) consistent with its polymerization at a lower temperature (7). The data for PVC-2 and 3 were presented in a previous paper (3); selected portions are used here for review and comparison.

Tetrahydrofuran (THF) used in fractionations and in GPC measurements was distilled under dry nitrogen using lithium aluminum hydride and stabilized with 0.025% 2,6-di-tert-butyl-p-cresol.

1,1,2,2-Tetrachloroethane (TCE) used for solution infrared measurements was dried using calcium chloride and distilled under vacuum at approximately 70°C.

B. FRACTIONATION

Prior to fractionation, polymer solutions were made up as 0.5 weight percent concentrations in THF. The solutions were then heated to 85°C under N_2 for four hours to disintegrate all molecular aggregates. Samples were fractionated using the THF-H_2O method as follows (3,8).

For each sample, the fractionation solution was placed in a vessel equipped with an agitator, a constant temperature coil, a non-solvent (H_2O) inlet, a nitrogen inlet, and a heavy phase withdrawal stopcock (3,8). Batch sizes of approximately 10 liters were processed in this type of equipment. The solution was brought to 25.0°C ($\pm$ 0.1°C) under N_2 and agitation and nonsolvent was added until a turbidity endpoint was reached. The solution was then allowed to "ripen" for several hours at which time the temperature was raised approximately 10°C to dissolve the precipitate. The solution was next cooled down to 25.0°C and the precipitate was allowed to settle out of solution; in all cases the settled precipitate was a gel. The precipitate was then withdrawn and dissolved in a small amount of THF, a film was cast from this solution, and traces of THF were removed from the film by vacuum drying at room temperature and a pressure of 25 microns. This procedure was carried out for each successive fraction until no more polymer could be precipitated out of solution. The procedure of collecting each fraction as a film was adopted merely as a convenience for storing the material prior to subsequent solution characterization steps.

C. CHARACTERIZATION

GPC Measurements: Gel permeation was carried out under ambient conditions with a Waters Associates GPC/ALC 301 Chromatograph using distilled, stabilized THF as the carrier solvent. The flow rate was 1 ml/min, the injection time was two minutes, and the sample concentration was 0.25 weight percent or less. A four-column system was used containing styragel with permeabilities of 7×10^6, 1.5×10^5, 1.5×10^4 and 5×10^3 Å. The columns were calibrated using narrow molecular weight distribution polystyrene standards obtained from Waters Associates. Axial dispersion was found to be minimal and it was therefore disregarded.

Following a procedure similar to that of Abdel-Alim and Hamielec (5), GPC scans of the fractions and whole polymer were run before and after heating the solutions to 85°C for 30 minutes in pressurized glass bottles.

It was noted that in most cases after the heat treatment the anomalous second, high molecular weight GPC curve had completely disappeared and further heating resulted in no additional change in the GPC curve. After heat treatment at 85°C all the solutions were colorless, suggesting that degradation did not occur.

GPC weight and number average molecular weights were calculated using the Q factor method with the value Q = 25.0 for PVC verified by comparison with standard PVC samples which were characterized by light scattering and osmometry.

D. IR SYNDIOTACTICITY MEASUREMENTS:

To determine relative syndiotacticities for various samples we used the solution infrared spectroscopic method of Germar, Hellwege and Johnsen (9) to determine values of S, the fraction of syndiotactic diads. Our technique is reviewed in reference 3. Statistical analyses of the infrared data show that the syndiotacticity values we obtained by this technique are reproducible to $\pm$ 1%.

Some difficulty was encountered in applying the IR syndiotacticity method to the first few fractions of PVC-5 because of solubility limitations in TCE. Fractions 1, 2 and 4 were partly insoluble, even when heated to 90°C for several hours. S values for these fractions were calculated by the following procedure.

As mentioned previously Figures 1-3 illustrate the difference between GPC chromatograms for unheated and heated samples of PVC-3 whole polymer and for the first two fractions. Figures 4 and 5 show the same for PVC-5, fractions 1 and 4. The hatched area

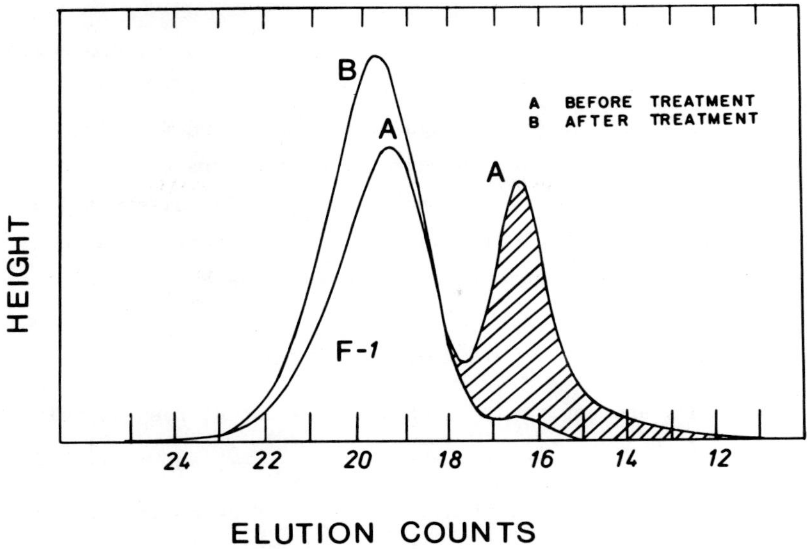

Figure 4: GPC, chromatograms for PVC-5, fraction 1 before (A) and after (B) heating solution; distribution initially is bimodal due to molecular aggregation. Note that aggregate peak does not completely disappear after heating.

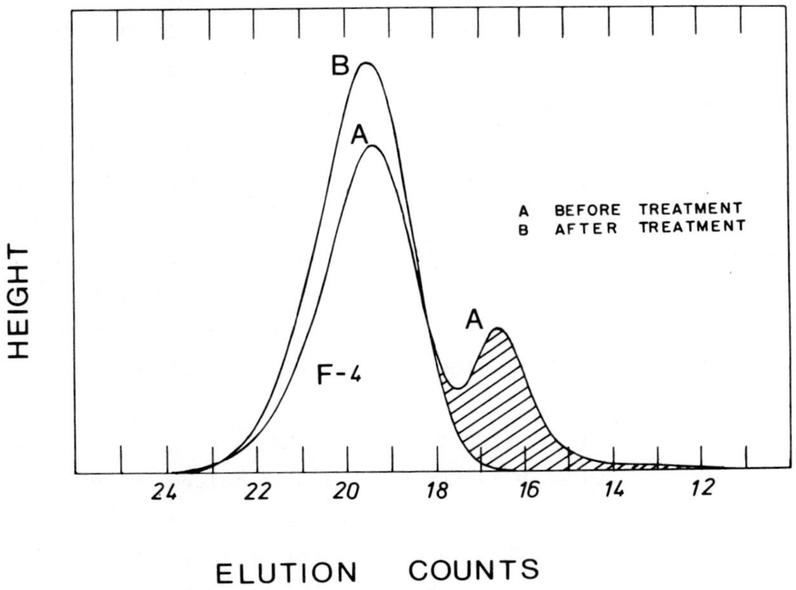

Figure 5: GPC chromatograms for PVC-5, fraction 4; bimodal distribution is also apparent in curve A.

represents an amount A, the high molecular volume fraction of the GPC chromatogram area due to aggregation (the fraction of the chromatogram area at high molecular volume which disappears after heat treatment). Abdel-Alim and Hamielec (5) correlated the fraction A with syndiotacticity, S, measured by NMR according to equation 1.

$$\left(\frac{1 - A}{A}\right) = k \left(\frac{1 - S}{S}\right)^h \qquad (1)$$

Here h and k are empirical constants, evaluated, respectively, from the slope and intercept of a plot of $\ln \left(\frac{1 - A}{A}\right)$ vs. $\ln \left(\frac{1 - S}{S}\right)$.

From the GPC values of A and those of S determined by IR for soluble fractions of PVC-5 and listed in Table 1, we calculated a value of h = 11.23 and k = exp(3.07). Equation 1 then gives the values of S listed for the various insoluble fractions of PVC-5 in Table 1.

In using the correlation, equation 1, we note the following. The linear relation between $\ln \left(\frac{1 - A}{A}\right)$ and $\ln \left(\frac{1 - S}{S}\right)$ holds well for PVC-5 fractions. The whole polymer, PVC-5 and all its fractions except fractions 1 and 2 were completely soluble in THF after heating, suggesting that the A values determined from the GPC curves are quite valid. Since small portions of fractions 1 and 2 were insoluble in THF (see Figure 4) even after heating, values of A for these two were estimated by eliminating the high molecular weight tail and assuming a smooth distribution curve. Fractions 5-30 were soluble in TCE. This indicates that the S values obtained by the IR method are accurate. In a previous paper we noted (3) that the S values which we obtain by the IR technique agree with those obtained by NMR for the same and similar polymers.

III. RESULTS AND DISCUSSION

While the evidence is strong that high molecular weight and/or stereoregularity contribute to enhance molecular aggregation in PVC, the results for PVC-5 and its fractions indicate that aggregation also occurs at short chain lengths and/or low stereoregularities (S values). In fact PVC-5 has a greater aggregation tendency than any of the other polymers we have studied.

In Figures 4 and 5 GPC chromatograms of unheated fractions 1 and 4, from PVC-5 are shown to have very large aggregate peaks. Essentially, the MWD curves for the fractions are truly bimodal. For the two fractions aggregate areas of the GPC curve are approximately A = 0.40 and A = 0.20. The whole polymer has molecular weight averages of $\overline{Mw}$ = 32,000, $\overline{Mn}$ = 15,300 (unheated) and $\overline{Mw}$ = 30,600, $\overline{Mn}$ = 12,000 (heated). The fractionation data in Table I show in addition that there are discrepancies between apparent $\overline{Mw}$ and true $\overline{Mw}$ values for the first five fractions of up to several hundred percent and there is a severe molecular weight inversion within the first few fractions. While the chains making up the aggregates in the first few fractions are in the 50,000 - 60,000 molecular weight range, it is obvious from the apparent and true number average molecular weights of the whole polymer that there is also a substantial amount of aggregation of short chains.

For this polymer the strong tendency for aggregation does not seem to be directly related to stereoregularity. The polymerization temperature of 50°C suggests an approximately atactic chain structure which is verified by the IR syndiotacticity data. Only fractions 1, 2 and 4 have S values over 50% and the values for these three are no greater than those for the high molecular weight fractions of PVC-2 (3), a higher molecular weight polymer which was also polymerized at 50°C. A comparison of the S values for all fractions of PVC-5 and PVC-2 in Figure 6 indicates that in general PVC-5 fractions have lower stereoregularities. We note, however, that the aggregation tendency for PVC-5 fractions as observed by GPC is far greater than for PVC-2 and even exceeds that for PVC-3, which has higher S values in the 60% range. A comparison of the GPC curves of Figures 2 and 4 demonstrate this. Moreover, the fact that PVC-5 aggregates are highly stable is suggested by the fact that autoclaving THF solutions of fractions 1 and 2 at 100°C for 72 hours does not disperse all of the aggre-

TABLE I

PVC-5 FRACTIONATION

Sample	Unheated		Heated		Weight %	A %	Syndiotacticity % (S x 100)
	$\overline{M}_w$	$\overline{M}_w/\overline{M}_n$	$\overline{M}_w$	$\overline{M}_w/\overline{M}_n$			
Original Whole Polymer	30,600	2.55	32,000	2.09	--	4.50	50.0
Fraction 1	330,900	7.23	58,500	2.07	2.73	37.00	55.6
2	227,300	6.31	54,900	1.91	1.25	20.00	53.8
3	66,200	2.48	49,600	1.91	1.3	4.80	50.2
4	180,000	4.92	54,000	1.90	1.0	20.70	53.8
5	74,900	2.01	67,100	1.80	1.0	2.70	48.9
6	66,300	1.56	66,700	1.52	4.14	0.80	46.1
7	57,600	1.45	57,900	1.45	9.04	0.80	46.1
8	52,600	1.40	51,500	1.41	4.46	1.90	48.9
9	48,800	1.38	48,400	1.37	2.41	0.70	45.8
10	46,600	1.40	46,300	1.36	3.02	1.10	46.8
11	43,000	1.37	43,300	1.34	3.26	1.80	47.9
12	41,000	1.35	40,600	1.35	2.87	1.70	47.8
13	39,000	1.32	38,400	1.32	3.55	1.70	47.8
14	37,100	1.29	36,500	1.29	2.59	1.40	47.4
15	34,700	1.32	34,100	1.30	3.8	1.50	47.5

16	32,100	1.28	32,400	1.28	3.27	1.50	47.5
17	31,100	1.29	31,200	1.29	1.9	0.80	46.1
18	29,800	1.26	29,500	1.26	3.65	1.30	47.2
19	27,700	1.25	27,100	1.26	3.62	1.90	48.1
20	24,900	1.23	25,400	1.24	2.99	1.50	47.5
21	24,100	1.23	24,300	1.24	3.18	1.50	47.5
22	22,800	1.23	23,000	1.24	1.27	1.90	48.1
23	21,700	1.23	21,700	1.23	4.34	1.30	47.2
24	20,400	1.23	20,100	1.22	2.3	1.20	47.0
25	18,600	1.22	18,300	1.23	4.81	1.60	47.7
26	17,000	1.22	16,700	1.23	3.45	2.50	48.7
27	15,200	1.21	15,700	1.22	3.37	2.50	48.7
28	13,900	1.21	14,000	1.21	3.48	0.70	45.8
29	12,700	1.22	13,200	1.23	2.17	3.60	49.5
30	11,300	1.22	11,500	1.23	2.27	1.60	47.7
					92.52%		

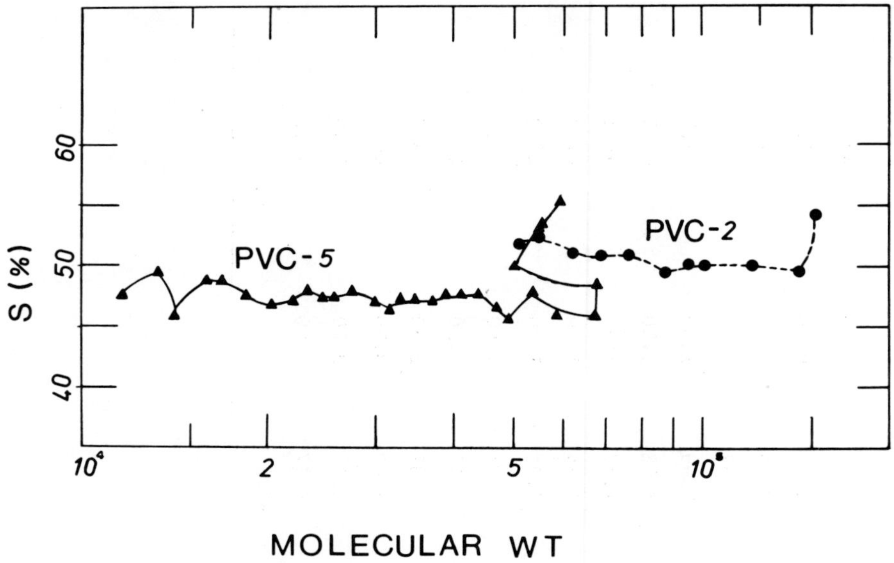

Figure 6: Syndiotacticity % S vs. molecular weight for fractions of PVC-5 compared to a higher molecular weight sample, PVC-2

gates. We have not encountered such a degree of difficulty in dispersing aggregates in any of the other PVC samples we have studied.

While one may argue the possibility (3) that long syndiotactic sequence lengths (but not high S values) are responsible for the high aggregation tendency we observe for short chain molecules, another explanation seems more appropriate especially if one considers the very low molecular weight aggregates responsible for the anomalous $\overline{M}n$ value for the whole polymer.

The high crystallization potential of low molecular weight PVC is quite well known. Böckman (10) suggests that this tendency is independent of tacticity and increases with decreasing molecular weight. Assuming that aggregates are a type of crystal structure, one would expect a high aggregation potential for PVC-5. Salovey and Gebauer (11) show that the aggregates have a diffuse x-ray diffraction pattern and suggest therefore that they do not contain true crystallites but nematic structures similar to those proposed for secondary crystallinity in bulk PVC by Juijn and co-workers (12). Such secondary crystals of PVC are thought to contain straight chain isotactic segments as well as syndiotactic segments and could form readily from highly mobile short chains during precipitation. Thus, according to this view atactic short chains should have a great potential for associating on precipitation. In this regard it is also worth noting that short chains are expected to have fewer branching and head to head type imperfections (10, 13).

IV. ACKNOWLEDGEMENTS

The authors wish to thank DR. E.A. Collins, B.F. Goodrich Chemical, Co., Avon Lake, Ohio for PVC Samples used in this study and for his useful comments and technical assistance.

Financial support for this work was provided in part by the General Electric Foundation, The Research Council and Department of Chemical and Nuclear Engineering Development Fund, University of Cincinnati.

V. REFERENCES

1. J.C. Yingst and R.P. Chartoff, "A Study of Stereoregularity in PVC using the Gel Permeation Chromatograph", paper presented at Pittsburg Seminar on Current Problems in Gel Permeation Chromatography, Pittsburgh, Pa. Oct. 31, 1973.
2. J.C. Yingst and R.P. Chartoff, J. Appl. Polymer Sci., 19, 1193 (1975).
3. R.P. Chartoff, J.C. Yingst, B.F. Brush, and S.K. Lo, paper presented at 2nd International Symposium on Poly(vinylchloride), Lyon, France, July, 1976; to be published in J. Macromol. Sci. Physics.
4. P. Kratochvil, M. Bohdanecky, K. Solc, M. Kolinsky, M. Ryska, and D. Lim, J. Polymer Sci, C-23, 9 (1968).

5. A.H. Abdel-Alim and A.E. Hamielec, J. Appl. Polymer Sci., 17, 3033 (1973).
6. G. Palma and M. Carenza J. Appl. Polymer Sci., 19, 2625 (1975).
7. A.H. Abdel-Alim, J. Appl. Polymer Sci., 19, 2179 (1975).
8. J.C. Yingst, M.S. Thesis, University of Cincinnati, Cincinnati, Ohio 45221, 1974.
9. V.H. Germar, K.H. Hellwege and U. Johnsen, Makromol. Chem., 60, 106 (1963).
10. O.C. Böckman, J. Polymer Sci., A-3, 3399 (1965).
11. R. Salovey and R.C. Gebauer, J. Appl. Polymer Sci., 17, 2811 (1973).
12. J.A. Juijn, J.H. Gisolf, and W.A. de Jong, Kolloid-Z., 251, 456 (1973).
13. A.L. Goff, I.P. Yakolev, V.M. Zhulin, and M.G. Gonikberg, Polymer Sci., USSR, A-11, 1487 (1969).

THE ELIMINATION OF WATER-BORNE IMPURITIES IN HIGH PERFORMANCE LIQUID CHROMATOGRAPHY

R. L. Sampson

Millipore Corporation
Bedford, Massachusetts

When water is used as a solvent, especially in reverse-phase gradient elution chromatography, impurities normally found in distilled water can cause serious problems.

Organic impurities, at best, will produce poor baselines. And often enough, they will generate spurious peaks.

Impurities can also cause substantial economic losses by shortening the life of costly columns.

Inorganics such as polyvalent ions and organic impurities poison ion-exchange columns and, together with inorganic impurities destroy aqueous gel permeation columns.

Bacteria, growing in stored distilled water will plug columns as will suspended particles.

Trace Enrichment

Reverse Phase Gradient Elution Chromatography is a unique method for analyzing a wastewater sample containing a broad spectrum of organic materials in trace amounts.

This technique, known as "trace enrichment" allows for large wastewater samples to be injected onto the LC column. Trace organic compounds are concentrated on columns and eluted (and thus detected) during a reverse phase gradient run.

High-purity water is a requirement for both purging the column prior to sample collection and as a mobile phase solvent for eluting the sample prior to UV detection.

One of the factors that has the greatest impact in this analytical method according to Waters is that essentially all detectors used routinely in liquid chromatography today are concentration sensitive. Therefore, it is important that the intrinsic sensitivity of these detectors be enhanced by increasing the amount of material initially injected into the system. This can be maximized in using a good, clean, consistent water, to clean the column from one analysis to another giving both short turnaround time and zero carry-over.

The Limitations of Distillation

Distilled water routinely contains dissolved organics, significant amounts of inorganics and a surprising number of suspended particles - even with double and triple distillation. Meticulous maintenance can keep these contaminants to a minimum. Because of factors inherent in the process itself, however, distilled water at its best varies too widely in quality to be acceptable for the sensitive techniques and column costs associated with liquid chromatography today.

Milli-Q Evaluation

Because of the problem encountered in doing sensitive trace enrichment techniques, and the laborious preparation of the water, it was decided to evaluate the Millipore High-Purity Water system for water preparation in high pressure liquid chromatography.

A study was undertaken to determine the significance in using Millipore's high purity water system for doing reverse phase gradient elution chromatography. The reason for the study was to determine from a practical standpoint whether or not the use of Milli-Q was significant in Waters' LC applications. All of the testing was done on site at Waters in their applications laboratory using their Milli-RO4/Milli-Q system and Waters' ALC/GPC Model 244 Liquid Chromatograph, with a 660 programmer.

The parameters used in this test were a comparison of a Milli-Q with reverse osmosis feed blank (160 ml load - Fig. 1), bottled distilled (40 ml load - Fig. 2), and waste water (Fig. 3). These were all run at two sensitivities, and at conditions which Waters personnel agreed were realistic. The 0.1 sensitivity scans are shown in this report. The gradients were run over 30 minutes, going from 0% acetonitrile (100% Milli-Q water) to 100% acetonitrile (0$ Milli-Q water). One can see in looking at these results

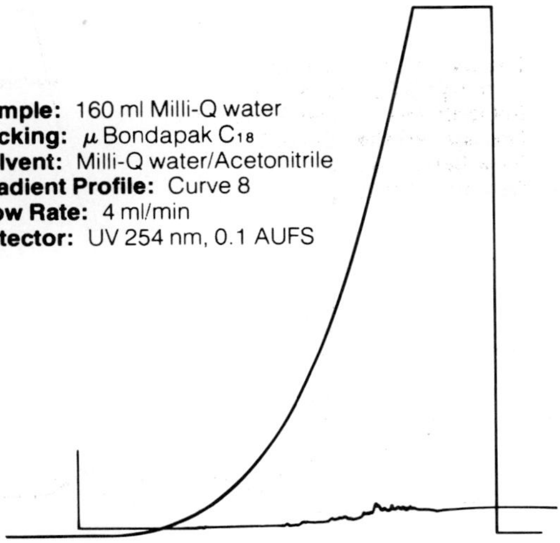

Figure 1. The blank from the Milli-Q System supply used to prepare the column and as the solvent for elution in Figures 1 and 2. Note that the sample size is four times greater than that in Figure 2.

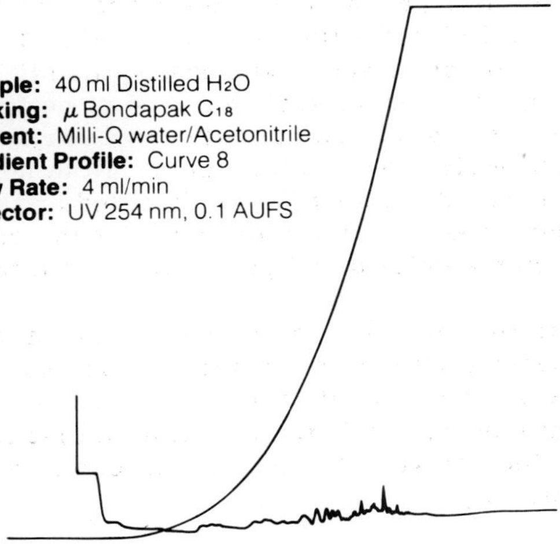

Figure 2. The blank from a distilled water supply. The interfering peaks make interpretation impossible. Note that the distilled water blank is one-fourth the size of the sample in Figure 3.

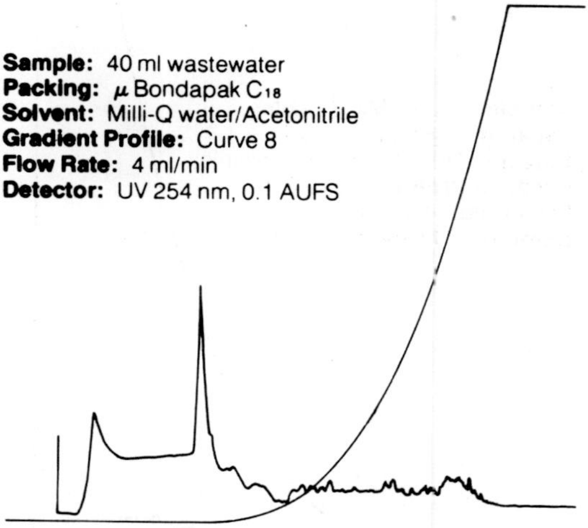

Figure 3. A wastewater sample using a Milli-Q System as the source for the elution solvent.

that the Milli-Q is significantly better. In the case of the Milli-Q there was a total of 160 ml due to the initial 10 minutes load at 4 ml/min (40 ml) plus the full 30 minute run at 4 ml/min (120 ml). This put the most stringent conditions on the Milli-Q water. It was agreed by both parties that this kind of performance by the Milli-Q would allow for consistently low background and would enhance trace enrichment applications. (Note here that the background contaminants as seen in the distilled and waste water figures, when used as a clean-up water for the column, could accumulate on the column and yield even worse results.)

In addition to the distilled water several other analyses were done with other water purification techniques such as distilled water through a carbon column with final filtration, Fig. 4, Central DI with RO pretreatment and Fig. 5, boiling of distilled water to remove low molecular weight organics that come over with the condensate, Fig. 6.

It was also found that when new expendables were placed in the Milli-Q system the first 30 to 40 liters had to be discarded in order to get the lowest background for the most sensitive analyses. This higher background was also found when the central system was regenerated and when the still was cleaned with acid because of scale formation.

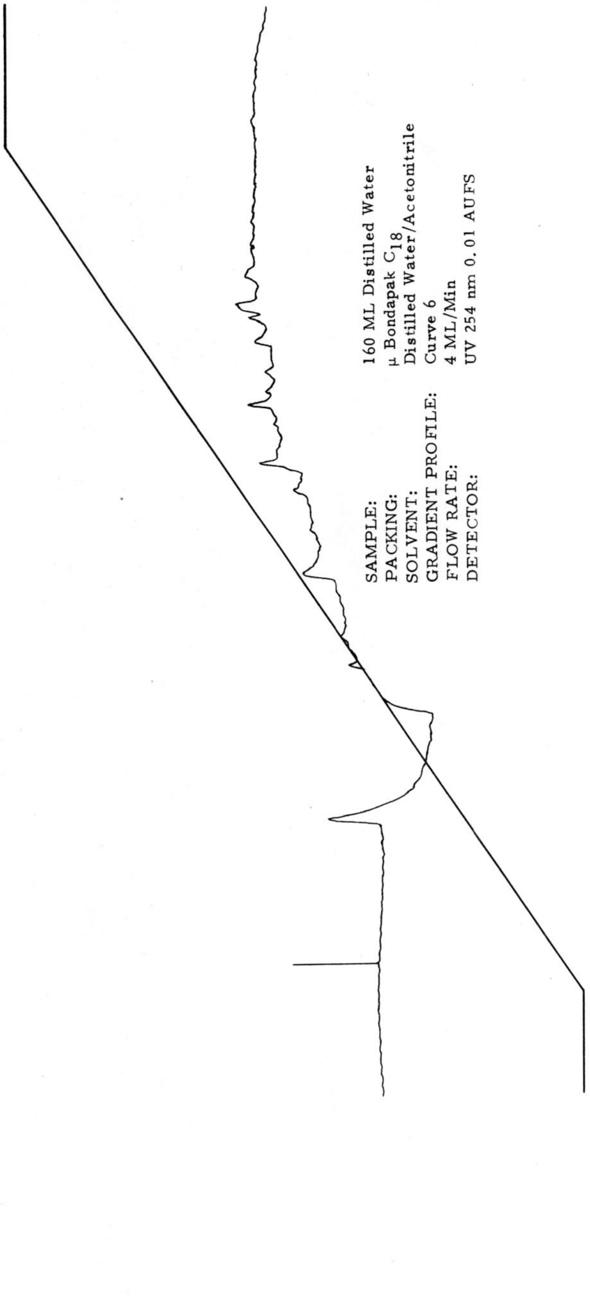

Figure 4. Fresh distilled water through a 30-inch carbon column.

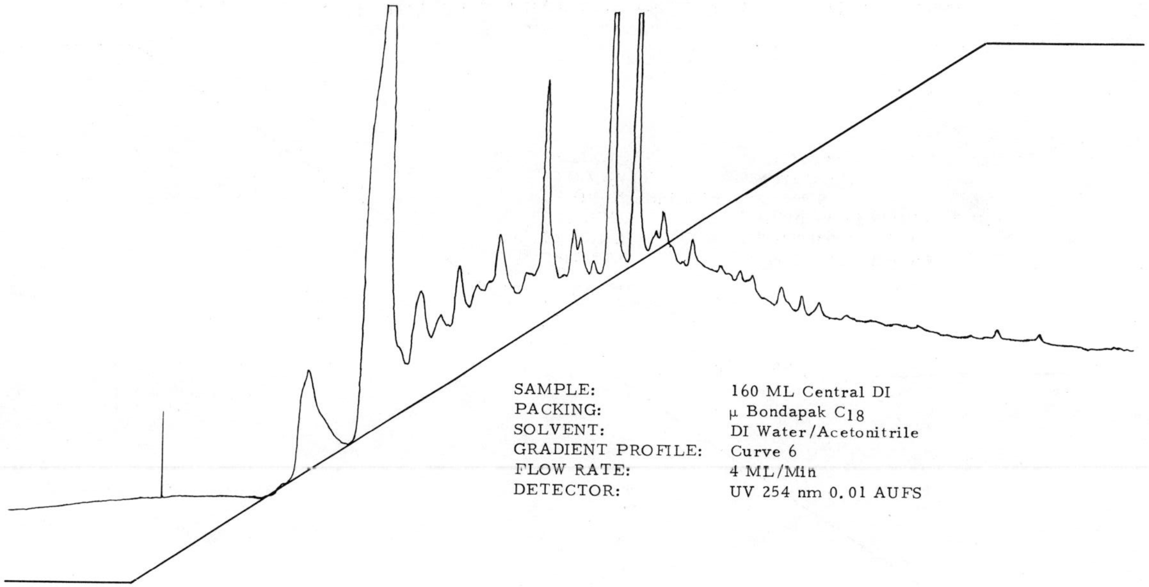

Figure 5. Central DI water with carbon pretreatment.

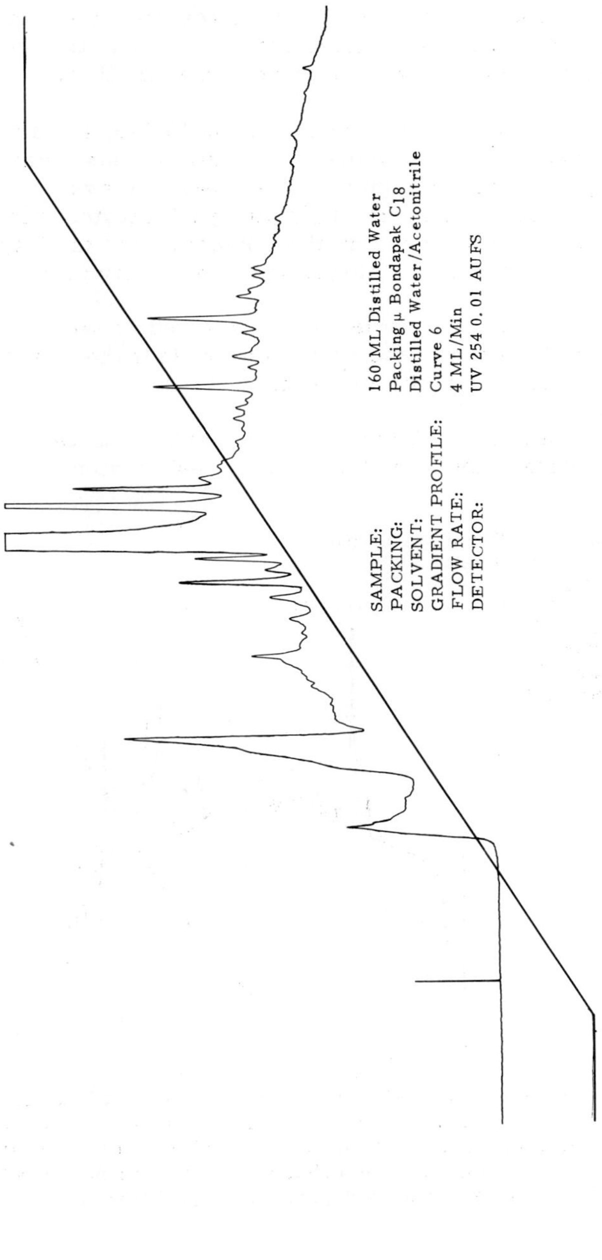

Figure 6. Fresh distilled water.

Applications

Once the technique for producing the best quality of water was determined, it was time to compare the results on an actual application. Biochemical separations are a most sensitive analyses and can be subject to any extraneous contamination.

Rapid separation and determination of the PTH-amino acid derivatives is a widespread application for reverse-phase gradient elution liquid chromatography. Water quality plays an important role. The column is prepared by purging with water, the sample is in an aqueous solution, and the solvents used to elute the sample are mixtures containing from ten to ninety percent water.

False peaks mask the material to be analyzed or, in certain cases, can shift the location of the derivative or limit the sample loading due to displacement of the base-line.

The three figures 7, 8, 9 that follow illustrate the value of a clean, consistent water supply such as the Milli-Q system.

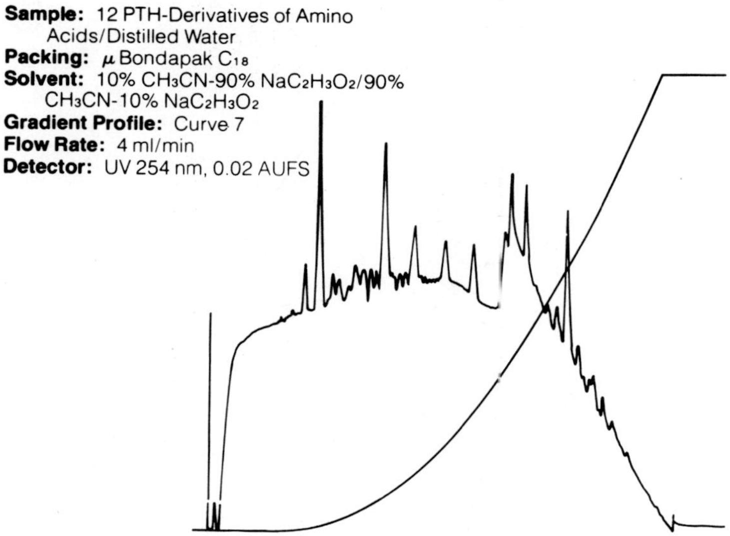
Sample: 12 PTH-Derivatives of Amino Acids/Distilled Water
Packing: μBondapak C_{18}
Solvent: 10% CH_3CN-90% $NaC_2H_3O_2$/90% CH_3CN-10% $NaC_2H_3O_2$
Gradient Profile: Curve 7
Flow Rate: 4 ml/min
Detector: UV 254 nm, 0.02 AUFS

Figure 7. A sample containing the same 12 derivatives as in Figure 4 but using a distilled water supply for preparing the column, buffers, solvents, and sample. Note that only 11 of the 12 derivatives are detectable. The background absorbance creates problems with sensitivity, limiting sample loads and peak interpretation.

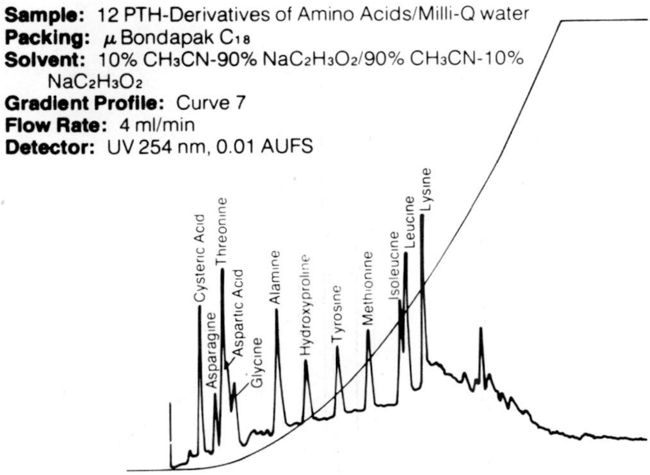

Figure 8. A sample containing 12 derivatives of amino acids using Milli-Q System reagent water to prepare the column, buffers, solvents, and sample. Note the sensitivity and ease of interpretation.

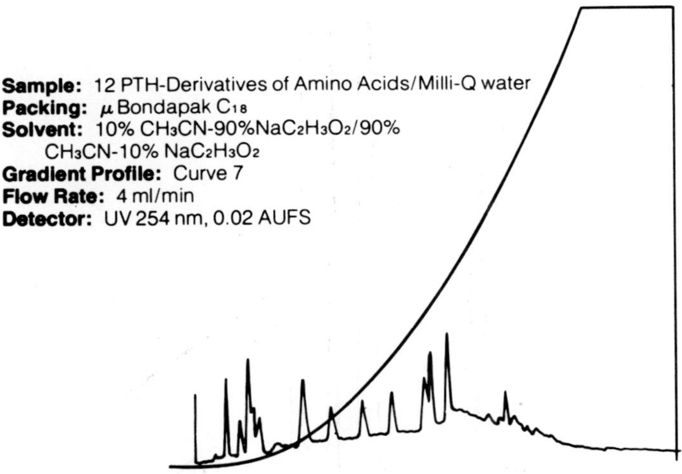

Figure 9. A sample containing the same 12 derivatives as in Figure 4 and using a Milli-Q System for all water requirements. This chromatogram was performed with the same sensitivity as with the distilled water (Figure 5) and illustrates the ease of interpretation even at this lesser sensitivity.

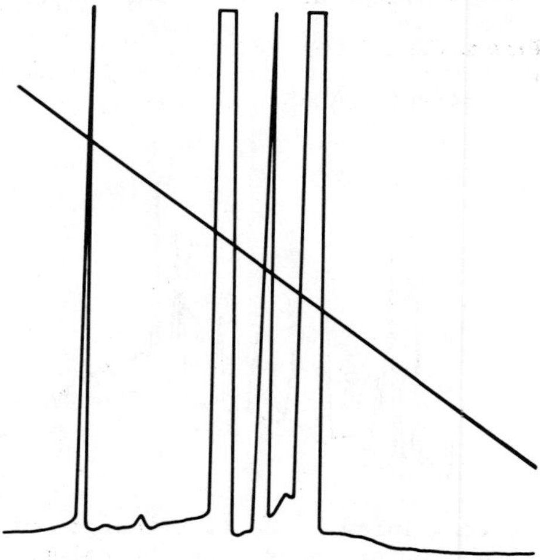

Figure 10.

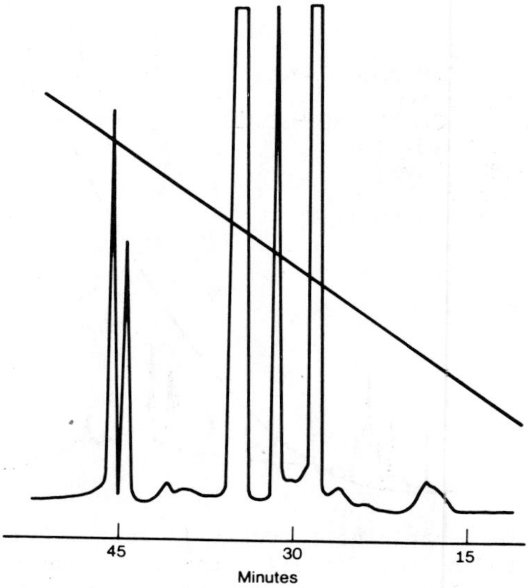

Figure 11.

Bottled Distilled Water

In a paper presented at the 27th Annual Pittsburgh Conference on Analytical Chemistry and Applied Spectroscopy and later reprinted in R/D in September, 1976, by Ms. Cheryl Creed of LCS Laboratories, Oakmont, Pennsylvania, it was shown that Liquid Chromatography simplifies isolating organics from water providing that water used as the primary solvent does not contribute other organic contaminants which would interfere with the assay.

This paper goes on to explain the tedious work that had to be performed to make the bottled distilled water useful as a solvent for use in the Water Associates Model 244 Liquid Chromatograph with UV detector and a μ Bondapak C18 column. Fig. 10 is a chromatogram of a 15 ml sample of bottled distilled water such as that purchased in polyethylene containers. These peaks correspond to a 7.5 ml prepared mixture of phlatalates, Fig. 11, indicating that there are plasticizers in this water. It can be readily seen that these peaks would probably interfere with any sensitive analysis that you might be preparing.

In conclusion it can be seen that a properly designed and maintained water system can enable you to do the most critical chromatographic separations without outside solvent interference. Because of its consistent high quality the Milli-Q system is also amenable to many other analytical procedures as well as high pressure liquid chromatography. One of these other applications is in Atomic Absorption Spectrophotometry. The Milli-Q system is recommended for use in graphite oven atomic absorption spectrophotometry by the major manufacturers of this equipment because of its extremely low ionic background, because they find that service calls are reduced by 50% when customers use the Milli-Q system.

38. Prechner, R., Panaris, R. and Benoit, H., Makromol. Chem. 156, 39-54 (1972). "Application of Gel Permeation Chromatography to the Study of Branched Polyethylenes." (in French)

39. Ram, A. and Miltz, J., J. Appl. Polym. Sci., 15, 2639-44 (1971). "New Method for Molecular Weight Distribution Determination in Branched Polymers."

40. Rogers, M.G., J. Appl. Polym. Sci., 16, 1953-8 (1972). "Structure of Epoxy Resins Using NMR and GPC Techniques."

41. Schultz, A.R., Eur. Polym. J., 6, 69-79 (1970). "Predicted Gel Permeation Behavior of Random Distribution Polymers Having Random Tri-or Tetra-Functional Branching."

42. Schultz, A.R., J. Polym. Sci., Part A-2, 10, 983-91 (1972). "Proposed GPC and Intrinsic Viscosity Analysis of Randomly Cross-linked Polymers Having Primary Distributions of the Schultz-Zimm Form."

43. Servotte, A. and DeBruille, R., Makromol. Chem., 176, 203-12 (1975). "Determination of Long-Chain Branching Distribution in Polyethylene by Combination Gel Permeation Chromatography and Viscometry."

44. Starck, P. and Kantola, P., Kem.-Kemi, 3, 100-5 (1976). "Determination of the Molecular Weight Distribution and Long-Chain Branching of Polyethylene Using GPC and a Remote Computer." (in Finnish)

45. Stockmayer, W.H. and Fixman, M., Ann. N.Y. Acad. Sci., 57, 334-52 (1953). "Dilute Solutions of Branched Polymers."

46. Strazielle, C., Pure Appl. Chem., 42, 615-25 (1975). "Molecular Characterization of Commercial Polymers."

47. Tung, L.H., J. Polym. Sci., Part A-2, 7, 47-55 (1969). "The Determination of Branching Distribution by Concurrent GPC and Sedimentation Velocity Experiments."

48. Tung, L.H., J. Polym. Sci., Part A-2, 9, 759-62 (1971). "Calculation of Long-Chain Branching Distribution from Combined GPC, Sedimentation, and Intrinsic Viscosity Experiments."

49. Tung, L.H. and Knight, G.W., J. Polym. Sci., Part A-2, 79, 1623-6 (1969). "GPC and Sedimentation (Flotation) Velocity Measurements on Linear and Branched Polyethylene."

50. Westerman, L. and Clark, J.C., J. Polym. Sci., Polym. Phys. Ed., 11, 559-69 (1973). "Low-Density Polyethylene. Variation of Branching Frequency with Molecular Weight and Its Influence on Molecular Weights as Determined by Gel Permeation Chromatography."

51. Wild, L., Ranganath, R. and Ryle, T., J. Polym. Sci., Part A-2, 9, 2137-50 (1971). "Structural Evaluation of Branched Polyethylene by Combined Use of GPC and Gradient-Elution Fractionation."

52. Williamson, G.R. and Cervenka, A., Eur. Polym. J., 10, 295-303 (1974). "Characterization of Low-Density Polyethylene by Gel Permeation Chromatography. II. The Drott Method Using Fractions."

53. Zimm, B.H. and Kilb, R.W., J. Polym. Sci., 37, 19-42 (1959). "Dynamics of Branched Polymer Molecules in Dilute Solution."

54. Zimm, B.H. and Stockmayer, W.H., J. Chem. Phys., 17, 1301-14 (1949). "The Dimensions of Chain Molecules Containing Branches and Rings."

AUTHOR INDEX

Authors' names are followed by page numbers and, in parentheses, reference numbers.

SUBJECT INDEX